BIRDS AT TRING RESERVOIRS

Rob Young, Jack Fearnside and David Russell

with additional research by Roy Hargreaves,
John Richardson and Johne Taylor

Illustrated by Jan Wilczur

Hertfordshire Natural History Society

25 YEARS ANNIVERSARY
HERTS BIRD CLUB
1971-1996

ISBN 0 9521685 1 0

First published in 1996 by the Hertfordshire Natural
History Society

Published with the support of British Waterways.

British Waterways

Printed in the UK by Admin Press Limited,
Watford, Hertfordshire

Table of Contents

Foreword

When canals were being built almost two centuries ago the canal builders had no idea of the wonderful investment they were making for future generations. Hard engineering structures were being created with only one purpose in mind, that being the transport of goods and the creation of commercial profit.

Time has changed this. The maturity of the waterways has increased the value of the original investment but this value is now judged in environmental terms rather than commercial terms. This change in the basis of valuation came about because of the way that the landscape and wildlife have adopted the waterways and reservoirs. What was once an intrusion is now established and part of our natural and working heritage.

Nowhere is this more obvious than around the Tring Reservoirs. They are not only used by British Waterways to supply water to the Grand Union Canal but they are also an important wildlife habitat accessible by many to appreciate and enjoy. This book will widen that appreciation.

Chris Mitchell
Waterway Manager, Grand Union Canal (South),
British Waterways

Introduction

The flora and fauna of the UK are amongst the best documented in the world. In excess of 60 million biological records exist including almost 40 million relating to birds (Department of the Environment 1995). Whilst local studies such as this can only be of limited influence, they can act as building blocks for national studies and play an important role in helping to preserve the unique nature of individual sites.

Tring Reservoirs forms one of the most important wetland sites in both Hertfordshire and Buckinghamshire. Between 1955 and 1979 the area was Hertfordshire's only National Nature Reserve and has been designated as a Site of Special Scientific Interest, or equivalent, since 1953.

There is a long history of ornithological recording at the reservoirs. Records in the 19th century generally comprised of birds that did not leave the area alive, and a large number of these were provided by that prolific collector Lord Walter Rothschild who was a regular visitor to the area between 1885 and 1920. Another long series of records was compiled by Charles Oldham between 1908 and 1936 and, following the discovery of Britain's first breeding Little Ringed Plovers in 1938, a large number of observers have regularly visited the site. These records were supplemented by observations carried out by staff of the British Trust for Ornithology (BTO) when offices were established at Beech Grove in Tring in 1963. The late Bob Spencer commenced ringing at the reservoirs in 1967 and his work developed as the forerunner of the Constant Effort Sites Scheme (CES) which was launched nationally by the BTO in 1986.

Records for Tring Reservoirs have been published as part of the annual county bird reports, in many books from Hartert & Jourdain (1920) to Sage (1959), Gladwin & Sage (1986) and Lack & Ferguson (1993), and in a separate publication, *The Birds of Tring Reservoirs* by Holdsworth *et al* (1978). This latter booklet is now out of print and only carries records up to the year of publication.

Historic ornithological highlights have included Britain's first Marsh Sandpiper in 1887, the first breeding of Black-necked Grebe in England during 1918 and the aforementioned first British nesting of Little Ringed Plover in 1938.

During January 1978, Tony Prater at the BTO began a systematic log of bird sightings at the reservoirs. This continued until the BTO's departure to Norfolk in 1991 when the log books were passed to the Herts Bird Club, a section of the Hertfordshire Natural History Society. Since 1993 the task of coordinating the records has been undertaken by the Friends of Tring Reservoirs in close liaison with the Bird Club.

There are a great many people who visit the reservoirs for recreational purposes either on a regular or occasional basis and it is an aim of this publication to supply a ready guide to the habitats and bird species they can expect to see. A second, and perhaps more critical aim, is to document the status and changes in the area's bird life to provide information against which future comparisons can be made.

Acknowledgements

This book would not exist without the observations submitted by both known and unknown observers to the relevant bird-recording organisations. It would also not have been possible to perform the detailed analysis required without the daily log and we are greatly indebted to Tony Prater for initiating the process and to John Marchant, the late Adrian Cawthorne and Steve Dudley for committing the time necessary to maintain the log prior to the departure of the BTO. The log is now maintained by the Friends of Tring Reservoirs.

In addition to the authors, much of the initial research for the species texts was performed by the small team of Roy Hargreaves, John Richardson and Johne Taylor and it was largely due to this combined effort that the project finally got off the ground after many months of discussion. John Marchant and Paul Clark commented on later drafts of the text, and we are grateful for the benefit of their local knowledge. We would also like to thank Trevor James for helpful comments on the introductory pages.

The publication has mainly been sponsored by the Hertfordshire Natural History Society as part of its continuing programme to document the county's natural history. The production of this book has also been aided by a generous grant from British Waterways who have the onerous task of balancing the national ecological importance of the reservoirs with the day to day operational issues under sometimes challenging economic pressures.

The research has drawn heavily on the annual bird reports of both the Herts and Buckinghamshire Bird Clubs. Through Johne Taylor, the Tring and Aylesbury Vale Ringing Groups provided detailed ringing records from the two CES ringing sites at Wilstone and Marsworth Reservoirs respectively. Historically, ringing data was compiled by the late Bob Spencer, Will Peach and Kevin Baker.

We would like to thank Jan Wilczur, who's illustrations enliven the text and give such evocative views of Tring Reservoirs. We are also grateful to *British Birds* magazine for allowing us to reproduce the historic portrait of nesting Black-necked Grebe taken by the late Oliver Pike. Other photographs were taken by John Marchant (Red-necked Grebe, Long-tailed Skua), the late Adrian Cawthorne (Shags), Graeme Pegram (Spotted Crake), Johne Taylor (Kingfisher, Cetti's and Savi's Warblers, Bearded Tit) and Cliff Tack (Paddyfield Warbler), and we grateful for the opportunity to reproduce their work.

It is hoped that this book will encourage regular and casual observers to continue submitting their records to the relevant organisations and we would urge all visitors to contribute in this way. Contact addresses for the respective organisations can be found in Appendix 4 on page 129. RAY, JDF, DHR

BLACK-NECKED GREBE.

A history of Tring Reservoirs

The Chiltern Hills had long provided an impediment to the expanding network of canals in southern England and the opening of the Grand Union Canal in 1805, via a cutting through the chalk hills (the 'Tring Cutting'), represented one of the major engineering feats of the early 19th century. An integral aspect of the scheme was the requirement for a vast supply of water to be stored at the highest point near Tring and effectively to let the canal run downhill in both directions. Initially the demand for water must have been somewhat underestimated as the first reservoir at Wilstone (built in 1802) was soon joined by further capacity at Marsworth (1806), Tringford (1816) and Startops End (1817). The current configuration was completed by two extensions of Wilstone Reservoir in 1836 and 1839. The section of canal at the highest point, stretching for about five kilometres between Bulbourne and Cow Roast, is known as the 'Tring Summit' and is approximately 120 metres above the canal's junction with the River Thames at Brentford, Middlesex.

Even after the demise of commercial traffic on the canal in the 1960s there are still up to 6000 boat movements per year over the summit with each boat navigating the elaborate series of locks and an average in excess of nine million litres of water being lost downhill each day. The pumping station at Tringford now controls all of the water flowing from the reservoirs. Water is transported via underground culverts and is pumped into the Wendover Arm at the pumping station. Built in 1818, the original operation was powered by an impressive steam engine, although a smaller diesel powered unit is now used. There was formerly a series of three pumping stations along the Wendover Arm but two of these had been closed by 1836. The arm was used to transport local goods to and from Wendover but it continually suffered from leakage problems. It was finally closed to traffic between Little Tring and Drayton Beauchamp in 1904 and now constitutes the feature referred to as the 'dry canal', above the south side of Wilstone Reservoir.

A flour mill has been present at New Mill since 1875. Initially it was situated alongside a windmill and boatyard and this location provided both power and transportation with produce carried by narrow-boat along the canal. The site is still used in the production of flour but all transport is now by road.

Habitat, conservation status and other wildlife

With the attractive backdrop of the Chiltern Hills, the reservoirs have great æsthetic appeal and, despite being man-made, they are well established and have taken on many of the features of more natural lakes. Surrounded by farmland and areas of semi-natural marsh and fen, the four reservoirs represent a habitat increasingly rare in south-east England. Other nearby water bodies, at College Lake, Pitstone No. 2 Quarry and, to a lesser degree, Weston Turville Reservoir, also act as a refuge for waterfowl during times of disturbance and therefore Tring Reservoirs should be considered as part of a wider complex of waters. This has been underlined by recent studies on the use of College Lake by ducks taking refuge from shooting at Wilstone and Tringford Reservoirs (Russell pers obs).

The national importance of the reservoirs has been formally recognised since 1952 and their notification as a Site of Special Scientific Interest was re-confirmed in 1987.

While the reservoirs are artificial in construction, they are fed by natural springs. Water-bodies situated on chalk or limestone are referred to as marl lakes and, in common with other such waters, the reservoirs exhibit generally clean and clear water. Wilstone Reservoir remains the best example of this, although Marsworth Reservoir in particular has suffered from pollution in the past. This has led to that reservoir becoming less clear and the consequent silting-up has speeded the natural succession process. The reedbed has become drier, encouraging willow carr encroachment and a reduction of the open water area as the reeds have spread.

Despite being affected by rainfall over the Chiltern area, the water levels within the reservoirs can change for other reasons. A major factor influencing water levels is the amount of boat traffic passing over the summit of the Grand Union Canal. Temporary adjustments can also be made to individual levels by British Waterways who may actively pump water between the reservoirs for operational reasons.

The tall fen associated with the shelving slopes of the reservoir edges is largely dominated by Common Reed with some areas of Reed Sweetgrass. The expanse of the reedbeds at Marsworth and Wilstone Reservoirs combine to form one of the largest areas of reeds in the region. This attracts a large colony of Reed Warblers in the summer and acts as a roosting refuge for flocks of Corn Buntings and huge numbers of Starlings in winter. Occasionally rarer birds are located in the reedbeds but, because of their extent and the lack of public access, many more must go undetected.

Plant-rich areas contain Hard Rush, Celery-leaved Buttercup, Bulrush and the locally uncommon Lesser Bulrush. Rushy Meadow represents an unusual community type in the county being dominated by Blunt-flowered Rush, with frequent Meadowsweet and Common Fleabane. A variety of sedges is present including Distant Sedge which is more abundant here than anywhere else in Hertfordshire. Other plants recorded here and rare elsewhere in Hertfordshire include Fen Bedstraw, Southern Marsh Orchid, Early Marsh Orchid and Bog Pimpernel, although the latter has not been recorded recently. Another speciality is Round-fruited Rush, one of its very few localities. The open water supports few plant species, although Canadian Waterweed, and its relative Nuttall's Pondweed, Rigid Hornwort and Spiked Water Milfoil are all abundant. The muddy margins of the reservoirs vary greatly in extent but do support Orange Foxtail and the nationally uncommon Mudwort.

The woodland at the southern side of Startops End and Marsworth Reservoirs is predominantly Ash and poplar. Since the demise of the tall Elm trees, Tringford Reservoir is edged to the west with mature Horse Chestnut and there is a stand of Common Lime near the pumping station. The wet woodland to the east has mainly Alder, poplars and Sycamore with an understorey of Elder and small, and somewhat diseased, Elms. At Wilstone Reservoir the dominant trees are poplars with much mature Elder and an old overgrown fruit orchard in the private area. Marginal scrub here consists of Osier and is one of the few county sites where Green-flowered Helleborine can be found. Hedgerows and areas of scrub, such as the dry canal, feature a mixture of Hawthorn, Elder and Blackthorn.

Most of the commoner butterflies can be encountered and it is also possible to find a few scarcer species, such as Essex Skipper, Holly Blue and Ringlet, particularly along the dry canal and in the relatively unimproved meadow behind the hide at Wilstone Reservoir. The Comma and Speckled Wood are two species that have increased over recent years and there have been several recent sightings of the attractive Marbled White which is now common in the adjacent Chilterns. Migrants like the Red Admiral and the very occasional Clouded Yellow also occur, but by far the most unusual occurrence to date was Hertfordshire's only record of the spectacular Monarch (or

Milkweed) on the causeway between Marsworth and Startops End Reservoirs in August 1947, a most unexpected vagrant, probably from North America (Sawford 1987).

That other popular group of insects, the dragonflies and damselflies, is also represented at the reservoirs by a relatively limited number of species when compared with other nearby sites in the Vale of Aylesbury. More notable species include Emperor Dragonfly, Black-tailed Skimmer, Four-spotted and Broad-bodied Chasers, Common Darter and Emerald Damselfly, and good numbers of Southern, Migrant and Brown Hawkers often patrol the area behind Wilstone Reservoir in late summer. In August 1995 at least 10 Yellow-winged Darters were found during an unprecedented national influx of this irregular migrant from continental Europe.

Studies of the *Coleoptera* of Tring Reservoirs earlier this century showed it to be a very rich site, and it still records the largest number of nationally scarce species for any one site in Hertfordshire, although there is a great need for repeat studies to see if this interest remains (James pers comm).

Vast numbers of insects hatch at the reservoirs and this ample food supply supports good numbers of bats. Species vary from the largest British bat (the Noctule) to the smallest (the Pipistrelle) and include Daubenton's Bat, a species reliant on water for feeding. Two Brandt's Bats were trapped at Wilstone Reservoir in 1975, the first Hertfordshire record. Other mammals are not well recorded at the reservoirs but Muntjac are being increasingly seen and another alien species, the Mink, appears to be becoming more numerous. Foxes are often reported in the area with cubs seen in most summers. Rabbits are common, particularly around Wilstone Reservoir, where both black and light fawn colour morphs can be seen amongst the usual grey-brown population. Odd cases of myxomatosis can still be found, but it is not clear when and what effect Rabbit Hæmorrhagic Disease may have if it spreads to the area from south-west England. Otters were known to have been present until the early 1970s and may well reappear if the encouraging increase in their population continues.

A guide around the reservoirs

ACCESS

The area is served by many footpaths across the surrounding farmland, along the canal towpaths and around the reservoirs. Visitors should remain on the marked footpaths and be careful not to damage or disturb any of the nearby areas. Car parking is available at Wilstone and Startops End Reservoirs which both have information boards and provide easy access to the footpath network.

FIGURE 1. TRING RESERVOIRS.

Members of the public should be aware that shooting is carried out on a limited numbers of days during the winter months.

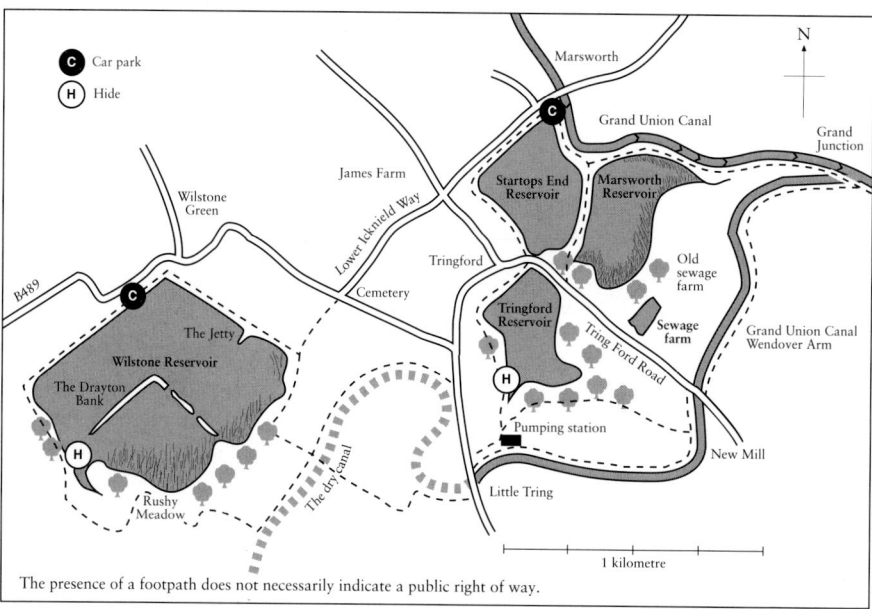

The presence of a footpath does not necessarily indicate a public right of way.

A QUICK GUIDE TO BIRDWATCHING AT THE RESERVOIRS

The following is intended as a brief introduction to some of the best areas to visit for those unfamiliar with the site and who have limited time. Activity will vary depending on the time of year, water level and time of day, although visibility is best looking west during the morning and east during the afternoon. Many good birds can and have been found away from the following sites and visitors are encouraged to explore the wider area from the numerous public footpaths and access points.

WILSTONE RESERVOIR

Open water	Diving ducks, gull roost, terns
Reservoir bank	Cormorant, heronry, Common Sandpiper, Grey Wagtail, crows flying to roost
Bird hide	Teal, Gadwall, Hobby, Water Rail, waders, Kingfisher, Starling roost
Jetty	Dabbling ducks including Wigeon and Pintail, Wheatear on the reservoir banks and adjacent farmland
Dry canal	Bullfinch, warblers and buntings
Woodland	Sparrowhawk, Tawny Owl, woodpeckers, tit species, Goldcrest

TRINGFORD RESERVOIR

Tring Ford Road	Great Crested Grebe, Kingfisher
Bird hide	Dabbling ducks including Teal, Kingfisher
Woodland	Tit species including Willow, Redpoll and Siskin in Alders, Stock Dove, woodpeckers

STARTOPS END RESERVOIR

Open water	Great Crested Grebe, diving ducks
Reservoir bank	Waders, wagtails
Paddock on north side of canal	Wagtails, Wheatear
Tring Ford Road	Waders in southern corner, Kingfisher
Woodland	Tit species, finches, Goldcrest

MARSWORTH RESERVOIR

Inlet at north-east corner	Teal, Snipe
Open water	Great Crested Grebe, Shoveler
Reedbed	Water Rail, Sedge Warbler, Reed Warbler, Reed Bunting, Corn Bunting roost

SEWAGE FARM

Lagoon	Teal, Green and Common Sandpiper, Snipe

OTHER SITES

Canals	Grey Heron
Hedgerows	Warblers including both whitethroats
James Farm	Corn Bunting, Collared Dove
Wilstone	
Cemetery area	Buntings, Mistle Thrush
Wendover Arm	Grey Heron, Grey Wagtail, Sedge Warbler
Farmland	Lapwing and Golden Plover, Little Owl, Skylark

A calendar of species

WINTER

The short days of winter can often bring rich rewards for those braving the elements. Indeed, extremes of weather will often result in unusual numbers of wildfowl or prompt large movements of winter thrushes, plovers or Skylarks.

The reedbed at Wilstone can hold a Bittern but, due to the relative inaccessibility of the area, the birds are best looked for at dusk when they can sometimes be seen flying over the reedbed. Bearded Tits can be present for long periods in winter but can effectively disappear for weeks between sightings in the dense reeds.

Waterfowl numbers are at their highest in winter with large numbers of Wigeon attracted to the site, especially to grain put down for the high population of captive-bred ducks released for shooting. There is, however, some evidence that other surface-feeding species are adversely effected by this activity with numbers of Shoveler and Coot in particular reducing over recent years (Figure 2).

The Mute Swan flock has increased in recent winters. Small numbers of Bewick's Swans visit most winters but the Whooper Swan is a true rarity. Other local rarities during the winter months include Scaup, both Slavonian and Black-necked Grebes, the few diver records and the occasional wreck of Shags. Relatively few scarce passerines are reported but Water Pipits are sometimes recorded when water levels are low.

Colder weather will often be the catalyst for the appearance of sawbills, but in milder conditions generally only Goosander will appear. Numbers of Goldeneye have increased in recent years with up to 20 birds present around the turn of the year. Very severe cold weather can disrupt wildfowl when large numbers are forced to leave

FIGURE 2.
WINTER
POPULATIONS
OF SELECTED
WILDFOWL
SPECIES,
1980/81-94/95.

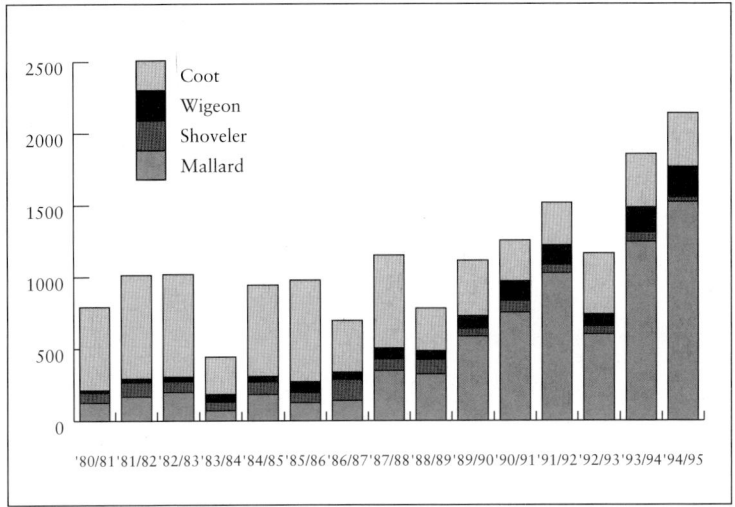

the site during periods of significant ice cover.

Roosting activity brings vast flocks of birds to the area from the middle of the afternoon. The largest mass of birds must be in the gull roost at Wilstone Reservoir but this gathering is rarely as spectacular as the sight of tens of thousands of Starlings whirling around the same reservoir before diving into the reedbed to roost as darkness falls. One of the largest Corn Bunting roosts in the region is in the reedbed at Marsworth Reservoir, although the future of this gathering must be in doubt given the dramatic fall in the population of this species. The presence of a crow roost nearby is indicated by the large numbers of Jackdaws, Carrion Crows and Rooks passing overhead in a north-westerly direction late each afternoon between October and the new year.

Sizeable flocks of small birds can be found during the winter months. These groups can contain all of the tit species resident at the reservoirs together with a few Goldcrests and the occasional Treecreeper. Chiffchaffs have overwintered increasingly in recent years and can be seen in the company of these flocks with up to five birds recorded in milder winters. These same ameliorating conditions may also have been responsible for the increasing occurrence of Blackcaps after the departure of the summering population, although there have apparently been few long-staying birds and most seen may relate to birds on passage.

**WILSTONE
RESERVOIR IN
EARLY SPRING.**

The coming of spring is first indicated by the early display of Great Crested Grebes at any time from the new year onwards. Grey Herons start to take up residence in the heronry at Wilstone during January and by the start of the spring they will already have young in the nest.

SPRING
The advent of spring brings longer days and the first signs of migration at the reservoirs. The first summer visitors to return are generally Sand Martins with records now often occurring during mid-March. When the habitat in the surrounding fields is suitable, Wheatears are usually seen soon after. Another traditional harbinger of spring, the Cuckoo, makes its presence evident from mid-April.

April and May usually produce small movements of waders, with records of Whimbrel being regular in this period. Bar-tailed Godwit, Turnstone and Sanderling can occasionally be seen with early mornings being especially favoured, though many occurrences involve birds which fly through without stopping.

From the middle of April the first Hobbies arrive. Increasing numbers of this agile falcon are now being recorded with a favourite time being during May. By this time records of both Marsh Harrier and Osprey have usually occurred. The latter often pass through rapidly, spending no more than a few minutes at the reservoirs, but Marsh Harriers have been recorded roosting in the reedbed at Wilstone Reservoir on several occasions.

Under the right conditions, large numbers of terns can pass through from the middle of April although numbers recorded are variable and very few are seen in some years. On the morning of 23 April 1994, after a spell of northerly weather, the wind swung around to the south-east and the overcast drizzly conditions produced a spectacular fall of Arctic, Black and Common Terns accompanied by Little Gulls and a Little Tern. Parties of Kittiwakes are recorded occasionally between March and May with flocks of up to 40 birds generally passing through quickly.

Little Owls are at their most obvious just before the trees break into leaf, particularly around Wilstone Reservoir. A number of territories are established in the area.

The reservoir banks attract a variety of wagtails during spring with White Wagtails often recorded with migrating Pied Wagtails and occasionally birds of the Blue-headed race join larger influxes of Yellow Wagtails.

Summer

Summer is traditionally the quietest time of the year for birdwatching with the northward migration ended and the breeding season in full swing. Depending on the spring weather, heron chicks may have left the nest, and ducks will have young on the water. During July a large number of Mallard are released for the winter shoot.

Common Terns first bred at the Reservoirs in 1993 and will hopefully do so again in the future. The nearby colony at College Lake continues to increase with large numbers of adult and juvenile birds now visiting the reservoirs from that site in late summer to feed.

The reedbeds harbour a large population of both Reed and Sedge Warblers and Reed Buntings are also numerous. All of the common warblers are present in the area with Blackcaps, Chiffchaffs and Willow Warblers in the woodland and both whitethroats in different hedgerow habitats.

The farmland surrounding the reservoirs supports several pairs of Skylark and Yellowhammer and occasionally a pair of Lapwings will attempt to breed.

Spotted Flycatcher numbers are somewhat variable but a few breeding pairs are to be found most years. The same cannot now be said of the Turtle Dove which is almost certainly extinct as a breeding bird at the reservoirs.

AUTUMN

Movements of Lapwings are often the first visible sign of southward migration although Green or Common Sandpipers can appear any time from mid-June. Water levels clearly have a major impact on the number of waders seen with some years producing only a few flyover records and others a good variety. When suitable areas of mud are exposed, the commonest species are Common and Green Sandpipers and Greenshank, the former two species often utilising the sewage farm lagoon. Other species, such as Little Stint, Dunlin, godwits, Redshank and Spotted Redshank, can occur and such rarities as Grey Phalarope, Pectoral Sandpiper and Long-billed Dowitcher have been recorded on occasion.

Autumn is the best time of year to watch Kingfishers, with family parties often present around the hides at both Tringford and Wilstone Reservoirs.

Duck numbers begin to build up during August and the site is important for large flocks of moulting ducks, particularly Tufted Duck and Pochard. Further influxes of the latter species have recently produced the occasional Ferruginous Duck. The number of dabbling ducks also increase and autumn is probably the best time to locate Pintail, although they are mostly in rather subtle eclipse plumage.

Rock Pipits are sometimes recorded from the reservoir banks and this season is also the best opportunity to see another reservoir rarity, the Nuthatch! Lack of suitable habitat means that this species is absent from the reservoirs despite its relative abundance in woodland within five kilometres of the site.

Systematic list

The following systematic list contains a summary of all published records for the site. For species which are or were designated as national or local rarities (current species marked with an asterisk), only records accepted by the bird clubs or the *British Birds* Rarities Committee have been included; for full references please refer to the appropriate annual bird reports or rarities reports respectively. Whilst this approach will have inevitably omitted some perfectly good records, it is in line with generally accepted practice and the authors would encourage any observer of omitted records to submit full details to the appropriate body to increase the accuracy of the archives. Records are included up to the end of 1995 with detailed analysis between 1980-95. A few 1996 reports are included.

Tring Reservoirs are host to significant breeding or wintering populations of a number of species of national importance (RSPB 1996). A suffix of '(R)' has been used to indicate birds on the 'red list', species which have reduced by more than 50 per cent as breeding species in the last 25 years. A suffix of '(A)' applies to birds on the 'amber list' indicating species which are suffering a moderately severe decline or whose UK populations are internationally important.

When interpreting the CES graphs given for some species, it should be remembered that the numbers of birds caught are related to the specific habitats where trapping occurs and may not necessarily indicate the level of populations throughout the area.

The details of some previously published records – Broad-billed Sandpiper, Great Snipe, Pomarine Skua, Marsh Warblers and Great Reed Warbler – would not be acceptable by today's standards. However, the Herts Bird Club Rare Birds Panel felt that any potential change in status should be considered as part of a future county-wide review and therefore these records are included in full. Conversely, a previously unspecified skua from 1919 was unanimously accepted by the panel as a Long-tailed Skua on remarkable photographic evidence. The authors believe that the 1887 Little Crake record is not sufficiently documented and should be disregarded. It is included, in square brackets, for continuity. A record of Rough-legged Buzzard (Sage 1959) was not within the boundaries covered by this work.

A table summarising the extreme dates for migrant species is included in Appendix 2 on page 125.

*Red-throated Diver *Gavia stellata*

Rare vagrant, 10 records.

The Red-throated Diver has always been a major rarity at Tring Reservoirs and it has not been recorded since 1975. Records have mainly fallen between late November and February with seven of the records occurring in January and February. Singles have been seen on 28 April 1975 and, quite exceptionally, on 3 July 1910.

*Black-throated Diver *Gavia arctica*

Rare vagrant, 10 records.

This species has been recorded on as many occasions as Red-throated Diver despite the fact that Black-throated Divers are much rarer on the nearest coastal waters. Birds generally appear during hard weather movements when their usual Baltic wintering grounds are frozen over.

At least eight have occurred between mid-December and mid-February with one in November, one in March and an extraordinary record of a diver believed to be this species on 14 May 1978. There were three present on 30-31 January 1937 but all other records refer to singles.

There has been only one record since 1980:

1983:   16 February, Startops End Res

*Great Northern Diver *Gavia immer*

Rare vagrant, 16 records.

This is the most frequent species of diver recorded at the reservoirs, although there are only 16 records to date. All have fallen between 31 October and 29 March, with 12 during the first two months of this period. Most birds have remained at the site for only a short period but the longest-staying bird remained for 25 days.

Two have been seen since 1980:

1982:   9-14 November, Wilstone Res
1987:   28-29 December, Wilstone Res

The 1987 bird took a Pike line but was freed and released.

Little Grebe *Tachybaptus ruficollis*

Resident, breeding species.

Often an elusive species, birds are resident at the reservoirs with fewer than five pairs breeding at Wilstone and Tringford Reservoirs. During the spring and summer months birds generally keep hidden in the waterside vegetation and their presence is often only disclosed by their high-pitched whinnying calls.

In the winter Little Grebes spend more time on the open water and the maximum recent count of 23 was made in December 1983. During cold weather birds can also be found on the canals.

This species was previously described as breeding 'fairly plentifully' and in the middle years of this century large flocks were noted in autumn, for example 76 in September 1937 and 79 in August 1952. Both breeding and non-breeding numbers have dropped in recent times with no counts greater than eight since 1987.

Great Crested Grebe *Podiceps cristatus*

Common resident.

The elaborate breeding display of this beautiful bird, once nearly hunted to extinction to satisfy the requirements of the millinery trade, is a familiar sight to anybody who visits the reservoirs in late winter and spring. It breeds with varying success every year and it is possible to see adults feeding young as late as November. Wintering counts can be impressive, with a maximum to date of around 95 birds in April 1995. There appears to have been a slow increase in numbers in recent years.

Sir Julian Huxley undertook his definitive study of the courtship rituals of the Great Crested Grebe at Tring Reservoirs, publishing his findings in 1914 (Holdsworth *et al* 1978). It is said that, at one time, all Great Crested Grebes in south-east England were descended from the last remaining pair at the reservoirs in 1866. By 1884, however, there were a staggering 75 pairs though numbers declined thereafter. Hayward (1947) recorded 20-30 pairs nesting in a normal year but the current population is now about 15 pairs.

*Red-necked Grebe *Podiceps grisegena*

Rare winter visitor and migrant.

Seventeen of the 24 records have occurred in the months between November to March, although the most recent records have been in migration periods and have included at least one bird in summer plumage. There is one old record of a bird staying over three months from 16 March to 22 June 1913.

With 19 records prior to 1980, the following have occurred since:

1980:   9-25 November, Marsworth Res
1982:   26 March

**RED-NECKED
GREBE AT
STARTOPS END
RESERVOIR,
26 FEBRUARY
1983.**

1982:   4 December until 27 February 1983, Startops End and Wilstone Res
1984:   11 September
1989:   9 & 26 April, Wilstone Res

*Slavonian Grebe *Podiceps auritus*

*Rare
winter
visitor and
migrant.*

This species continues to be unpredictable with 17 records between 1884 and 1980, mostly in the periods of October-November and February-April. The only summer records on 27 May and 26 June 1972, were presumed to be the same bird.

Since 1980 there have been six records involving seven individuals, all appearing in the period late October-early December except for one in April 1991.

*Black-necked Grebe *Podiceps nigricollis*

*Scarce
winter
visitor and
migrant.*

Prior to 1918 there were only seven or eight records of this small grebe but during that year three pairs nested at Tring Reservoirs, the first confirmed breeding in England. Proven or probable breeding also occurred on several occasions until 1928, and birds summered for the two following years. There were 15 further records from 1958 to 1980 mainly during spring and autumn passage periods.

In 1991 two birds seen on 12 May were displaying but had gone by the following day. With sporadic breeding occurring in the region, further breeding attempts may be possible although the increased

THE LATE
OLIVER PIKE'S
HISTORIC
PORTRAIT OF
BLACK-NECKED
GREBE
BREEDING AT
TRING
RESERVOIRS IN
MAY 1919.

disturbance at the reservoirs in recent years would tend to count against this, since Black-necked Grebes prefer undisturbed conditions for successful breeding.

The period since 1980 has seen 16 records with birds reported between mid-April and June and between the end of July and the end of November, with single January and February records. Exceptionally, two first-winter birds remained from 22 October 1994 until mid-November with one remaining until 27 December. All recent records have been from Wilstone Reservoir.

*Fulmar *Fulmarus glacialis*

Rare vagrant, one record.

1959:   female, 19-20 September, Wilstone Res

The occurrence of a Fulmar at one of the most inland stretches of water in Britain seems surprising although inland records are not unprecedented with five other Hertfordshire and five Buckinghamshire records. This bird was picked up in a moribund condition on its second day and was subsequently examined at the Tring Museum.

A record of a further bird in April 1968 appears in Holdsworth *et al* (1978) but no details have been published elsewhere.

*Manx Shearwater *Puffinus puffinus*

Rare vagrant, one record.

1985:   8-10 September, Wilstone Res

The only record relates to a three-day stay of this pelagic species and is remarkable considering this species' usual habit of moving on quickly from inland sites. This bird was first seen on the canal and later moved to Wilstone Reservoir. Early September is the most likely time for inland records with the dispersal of inexperienced juveniles away from west coast breeding sites, especially after gales.

*Storm Petrel *Hydrobates pelagicus*

Rare vagrant, two records.

1963:   18 November
1990:   22 October, Startops End Res

This species is generally less prone to being 'wrecked' after autumn gales than are Leach's Petrels and is consequently usually much rarer inland.

*Leach's Petrel *Oceanodroma leucorhoa*

Rare vagrant, three records.

This species has a reputation for being seen on inland waters after autumn gales for no more than one day. Most of the Tring Reservoirs records have followed this pattern.

1964:	22 November
1977:	15 November, Startops End Res
1990:	7 September, Wilstone Res

Interestingly the 1990 bird turned up the day after strong winds when the weather was sunny and calm, providing the very rare chance to see one settled on calm water.

*Gannet *Morus bassanus*

Rare vagrant, one record.

1990:   adult, 28 October, Wilstone Res

This bird appeared late in the afternoon and caused the roosting gulls to take flight while it circled round over Wilstone Reservoir. It made tentative attempts to settle on the water, before flying off to the south-west into Buckinghamshire.

In addition a bird was found in Wendover Woods on 4 December 1976 and released at the reservoirs.

Cormorant *Phalacrocorax carbo*

Frequent visitor, seen in all months.

The first documented records of Cormorants at Tring Reservoirs appear to relate to two or more birds during 1900 (Hartert & Jourdain 1920). Initial sightings were concentrated during the winter and migration months but recently numbers have increased and odd birds can now be seen throughout the summer with up to 53 birds at the Wilstone Reservoir roost during the winter months. They are at their most visible when perched in trees in the middle of Wilstone Reservoir or when drying their wings on other exposed perches.

Mid-summer records have only become regular since about 1989 and an immature bird started building a nest in early 1993. A small number of inland colonies have been established in south-east England in recent years and future breeding at the reservoirs must be a possibility (Sellers pers comm).

FIGURE 3.
AVERAGE
COMBINED
MONTHLY
MAXIMA OF
CORMORANT,
1980-95.

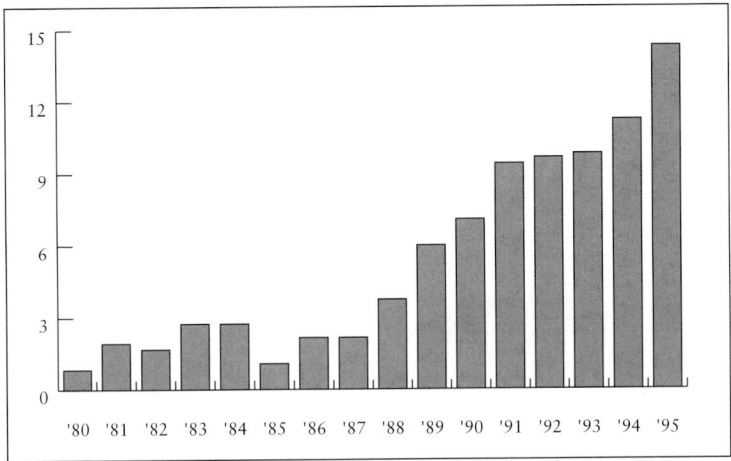

FIGURE 4.
AVERAGE
MAXIMA OF
CORMORANT
BY MONTH
(1980-95).

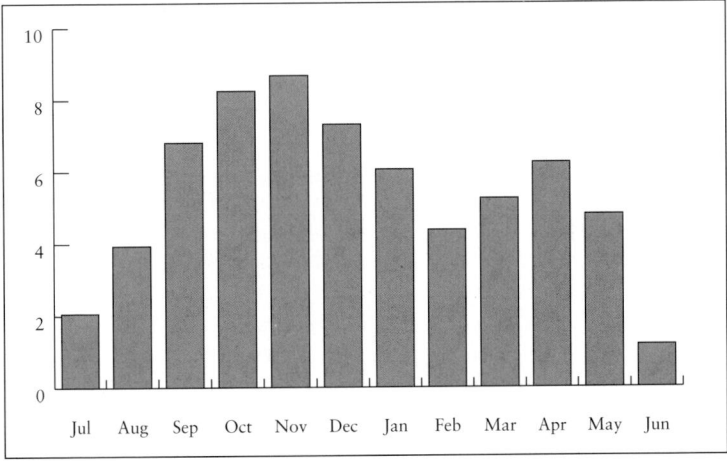

Shag *Phalacrocorax aristotelis*

Scarce winter visitor and migrant.

There have been records in eight years since 1980 with some indication of an increase in occurrence. While these records occupy a wide period outside the breeding season there is a pronounced peak in January and February. This is largely attributable to two 'wrecks', mostly involving juvenile birds, in 1984 and 1991 following periods of north-easterly winds. The largest count is of 28 juveniles in March 1962.

A 'WRECK' OF
NINE SHAGS AT
TRINGFORD
RESERVOIR ON
9 SEPTEMBER
1984.

Two individuals found dead after the 1984 influx had been ringed as nestlings on the Isle of May, Fife and the Farne Islands, Northumberland.

Bittern *Botaurus stellaris* (R)

Scarce winter visitor. The Bittern is an elusive and much sought-after visitor that has been recorded in all but two years since 1980. Even when birds are present, however, there can be long gaps between sightings which makes it difficult to establish the exact number of individuals involved. It is usually only in hard winters, when their usual feeding areas become frozen, that they are easier to see.

While winter occurrences peak in January, which are usually taken to be continental birds, there is a wide spread of dates between July and April. One unusual feature is a series of early autumn occurrences, with one record in July, four in August and one in September which may be birds of British origin.

The reedbeds at Wilstone and Marsworth Reservoirs are the best sites and the most successful strategy to see them is to scan the reedbeds just after sunset during the winter months.

In historic times a pair bred at Wilstone Reservoir in 1849 when four eggs were taken (Wolley 1853). This event seems unlikely to be repeated, although birds were present in the breeding season in 1927-29.

*Little Bittern *Ixobrychus minutus*

Rare vagrant, one record.

1968:   male, 17 August, Wilstone Res

There have been five other Hertfordshire and five Buckinghamshire records.

*Little Egret *Egretta garzetta*

Rare vagrant, three records.

Two records occurred during the invasion in 1989 when over 130 individuals were recorded in Britain; however, an additional bird was seen on 27 August at Wilstone Reservoir but details were not submitted to the *British Birds* Rarities Committee.

1989:   12-13 May, Wilstone Res
1989:   two, 20 September, Wilstone Res
1995:   adult, 13-24 August, Wilstone and Marsworth Res

The two 1989 birds flew into Buckinghamshire, providing that county with its first record (Lack & Ferguson 1993). Up to 1994 there have been six other Hertfordshire and four subsequent Buckinghamshire records. There can be little doubt that there will be more records in the future as this species continues its spectacular increase in Britain.

Grey Heron *Ardea cinerea*

Common resident.

This species is ubiquitous around the reservoirs and the activities in the heronry, built amongst the bushes along the centre bank of Wilstone Reservoir, are a familiar sight to any regular spring visitor.

During the period from 1956 until 1985, Grey Herons bred at Marsworth Reservoir in low Hawthorns and latterly in the reedbed but, possibly due to disturbance, the heronry moved to the Drayton Bank at Wilstone Reservoir. In recent years in excess of 30 pairs have raised over 50 young annually, making this the second largest colony in Hertfordshire. As with other colonies, there were noticeable increases in breeding numbers in 1989 and from 1994 onwards. These recent population increases may be due in part to the lack of hard winters which can cause severe reductions in numbers.

At any time of the year, Grey Herons can be seen hunting in the surrounding fields, on the reservoir banks or along the canal and will often afford excellent views.

FIGURE 5.
PAIRS OF GREY
HERONS AT
WILSTONE
RESERVOIR
HERONRY,
1985-95.

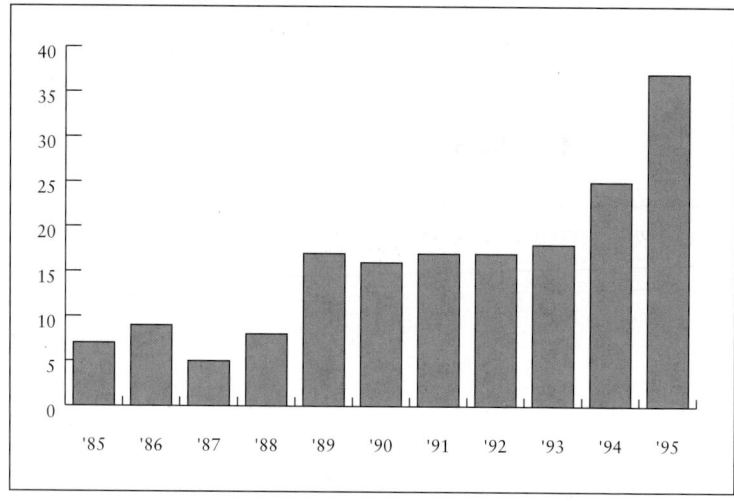

*Purple Heron *Ardea purpurea*

Rare vagrant, four records.

1958:	immature, 30 July-15 August, Wilstone Res
1967:	immature, 3 August-8 September, sewage farm field
1972:	adult, 7 May
1978:	juvenile, 30 August, Tringford Res

There have been four other records in Hertfordshire and two in Buckinghamshire.

*Glossy Ibis *Plegadis falcinellus*

Rare vagrant, one record.

1826:	one shot in October

The Glossy Ibis has declined as a European breeding species and the chances of a wild individual turning up again would seem to be remote. There have been two other Hertfordshire records and one in Buckinghamshire, by the canal at nearby Halton in 1886.

*Spoonbill *Platalea leucorodia*

Rare vagrant, three records.

Although this species is easy to identify, the recent occurrence of an escaped African Spoonbill at the reservoirs means that caution must be exercised and anybody finding a spoonbill should note the extent and colouration of the bird's exposed facial skin.

1947:   9 May, Wilstone Res
1957:   27 October, Tringford Res
1982:   two, 5 June, flew over Wilstone Res

Mute Swan *Cygnus olor*

Common resident.

A flock of non-breeding birds generally remains throughout the summer while between two and four pairs usually attempt to breed with mixed success.

Between August and November a significant influx is seen, resulting in the highest numbers of the year. During the new year this flock slowly disperses as breeding territories are established. The largest winter count to date is in excess of 140 birds during late 1995.

Numbers of this species are showing a strong recovery since the banning of lead fishing weights in 1987 (Gibbons *et al* 1993). This increase in population has resulted in more noticeable movements; observations of colour-ringed Mute Swans show that some of the birds wintering at Tring Reservoirs originate from the Thames valley and Oxfordshire with odd birds from as far afield as Worcestershire.

OCCASIONALLY A BEWICK'S SWAN CAN BE FOUND AMONGST THE FLOCKS OF MUTE SWANS.

*Bewick's Swan *Cygnus columbianus*

Scarce passage migrant and winter visitor.

Bewick's Swans are seen on the larger waters in most years at some point between October and March, with occurrences slightly more frequent during the mid-winter months. The largest flocks are generally seen during passage periods and spells of hard weather when up to 30 birds have been recorded (5 January 1985).

Migrant parties exceeding 10 birds usually pass through quickly, while the typical duration of a stay is less than three days. Occasionally smaller groups remain for longer periods, mainly during December and January, and particularly when water levels are low. This was certainly true during the winter of 1979/80 when two separate parties of three remained for about three weeks. Exceptionally, a first-summer bird was present during May 1996.

*Whooper Swan *Cygnus cygnus*

Rare vagrant, nine winter records.

Less frequently seen than the smaller 'winter swan', there are nine documented records of small parties of Whooper Swans visiting the larger waters, the most being 12 over Marsworth Reservoir in March 1960. All the records have fallen between December and March with the following reports since 1980:

1985:	two, 9 February, Wilstone Res
1985:	three, 17 March, Wilstone Res
1987:	three, 11 January, Wilstone Res

*Pink-footed Goose *Anser brachyrhynchus*

Rare vagrant and feral visitor.

A flock of 73 birds was seen over the sewage farm during a period of hard weather on 28 January 1979, although skeins of unidentified 'grey geese' passing over are sometimes thought to be this species.

In addition, there are a number of records of single birds or small parties and the majority of these presumably relate to feral or escaped birds.

*White-fronted Goose *Anser albifrons*

Rare vagrant and feral visitor.

On rare occasions parties are reported passing over between December and March and very rarely they have alighted. Records of single birds probably relate to escaped or feral birds.

There are about eight records from the reservoirs prior to 1980. Subsequent records are of single birds and the following of flocks which could relate to hard weather movements, and spring passage from wintering grounds in the south and west of England:

1987:	four adults and three immature birds, 18-19 January, Wilstone Res
1987:	33, 6 December, Wilstone Res
1990:	120, 3 March, over Wilstone Res, the highest number recorded

There was also a single bird at Wilstone Reservoir on 27-28 January 1996 at a time when many other birds were reported inland.

Greylag Goose *Anser anser*

Feral visitor. Greylag Geese were formerly common visitors between April and November with a maximum flock of 47 in September 1977. Between one and three pairs regularly nested but the last breeding record of that period was in 1978. In recent times records had diminished almost completely with only odd birds visiting the reservoirs, apart from a flock of 26 on 1 September 1993, and with no particular pattern to their occurrences until 1995. During that year a number of birds arrived at the reservoirs during the spring months after two birds had over-wintered with the local Canada Goose flock. Subsequently two pairs hatched goslings in 1996.

Canada Goose *Branta canadensis*

Breeding resident and abundant winter visitor. Canada Geese have been recorded at Tring Reservoirs since the 1940s and first nested in 1970 with up to five pairs currently breeding in the area. These birds are joined by a large number of visitors arriving during September and October, the period during which the post-breeding moult is undertaken, but these numbers reduce after November. The largest recorded flock to date is of 516 birds in October 1990.

*Brent Goose *Branta bernicla*

Rare vagrant, eight records. All records fall during the period November to March with the four November and the two March sightings possibly relating to passage movements. In recent years the frequency of records has grown and this increase seems to reflect the continuing build-up in numbers of Brent Geese wintering in southern Britain. Whilst there appears to be regular overland movements – sometimes involving large flocks – through the London area to these south coast wintering grounds, there does not appear to be any substantial evidence of this at Tring Reservoirs.

There have been six records since 1980 between early November and late February. A flock of 26, the largest recorded to date, was seen over Wilstone Reservoir on 3 January 1996 but the other records all involved between one and four birds.

Egyptian Goose *Alopochen ægyptiacus*

Rare feral visitor. Most occurrences of Egyptian Geese at Tring Reservoirs probably relate to birds which have recently escaped from captivity, although some records may refer to wandering feral birds now established in the wild and breeding in eastern England. Records since 1980 are as follows:

1985:	two, 12 January, Wilstone Res
1990:	16 March, Wilstone Res
1990:	an adult and two juveniles, 11-14 September, Startops End Res
1991:	14-15 December, Wilstone Res
1993:	five, 3 October, Wilstone Res, though at least one bird showed some suggestions of hybridisation with an unknown species

Shelduck *Tadorna tadorna*

Scarce migrant and winter visitor. Shelducks have arrived at the reservoirs in all months except June and July. Analysis of the records shows slight peaks during the periods March to May and September to October. Usually only one or two birds are involved but at least eight juveniles were reported during August 1989 and a flock of 12 birds was present on 24 March 1996. Often the birds prefer the larger area of Wilstone Reservoir but individuals or pairs were recorded, mainly at the sewage works, during the summer months between 1984 and 1986.

Some Shelducks recorded may be of captive stock, for example a long staying individual at Tringford and Startops End Reservoirs from August 1991 until May 1993.

Mandarin *Aix galericulata*

Scarce, mainly autumn visitor. With the increasing British population, and the establishment of this bird as a breeding species in Hertfordshire, the records are beginning to show a clear pattern. The first record at Tring Reservoirs was in October 1971. There were only three occurrences between 1980 and 1989 but records have subsequently been annual with a tendency for birds to occur between August and November or in January.

The largest party was a group of four males and a female on 9 September 1990. Mandarins usually prefer the secluded end of Tringford Reservoir.

Some records may still relate to birds of recent captive origin.

Wigeon *Anas penelope* (A)

Abundant winter visitor and passage migrant.

The first returning birds usually appear towards the end of August with numbers building up over the following two months and peaking between December and January. Counts of this species are variable with, for example, January maxima varying from 583 (1996) to just 15 birds (1989) in recent years, due to cold and mild weather respectively.

Passage birds pass through especially during late September-October and February-March when numbers are particularly

FIGURE 6. AVERAGE COMBINED MONTHLY MAXIMA OF WIGEON, 1980-95.

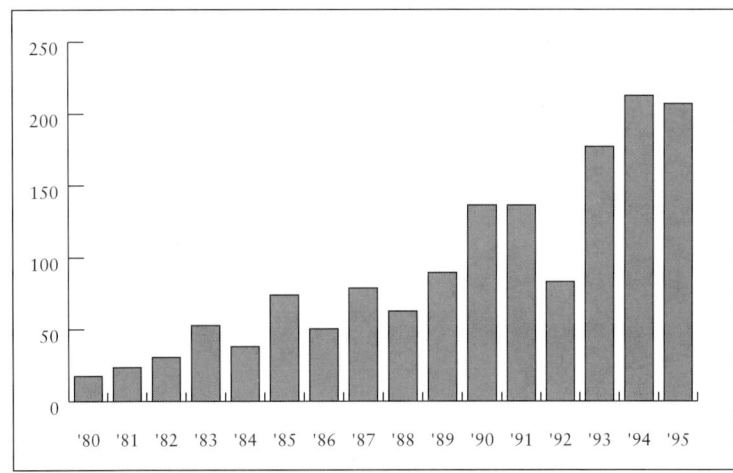

FIGURE 7. AVERAGE MAXIMA OF WIGEON BY MONTH (1980-95).

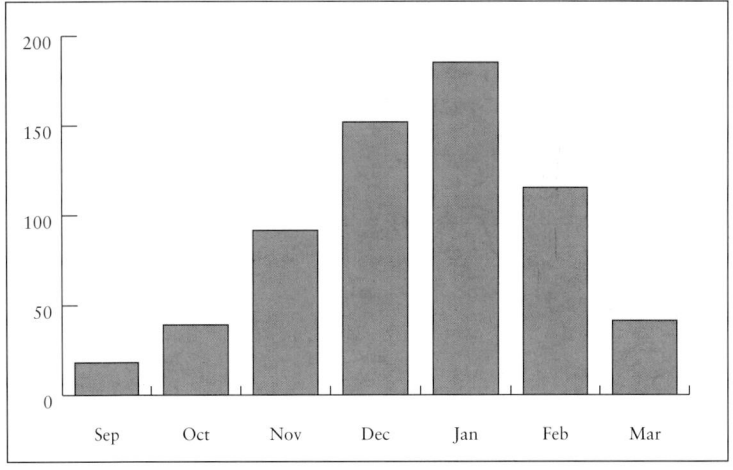

unpredictable. Occasionally odd birds remain into or visit during the summer months. Wilstone Reservoir is always the birds' preferred site where they appear to take advantage of the piles of grain put out for released Mallards.

*American Wigeon *Anas americana*

Rare vagrant and feral visitor.

1971:	male, 26 January, Marsworth Res, was the first accepted Hertfordshire record
1986:	first-winter male, 12 October until 8 February 1987, Tringford Res. No subsequent records of this individual were submitted to the *British Birds* Rarities Committee although it is believed that the bird remained past the latter date
1988:	first-winter male, 2-9 October, Tringford Res

Records of this species are frequently tainted with the possibility of being escapes from captivity. The 1986 bird arrived the day after a Pectoral Sandpiper at the sewage farm.

Gadwall *Anas strepera* (A)

Common, breeding in small numbers.

This subtly attired duck first bred in Hertfordshire in 1928 at Tring Reservoirs but then not again at this site until 1984. Recently up to 10 pairs have bred, although the numbers do fluctuate from year to year.

After the breeding season there is a build up of birds, most obvious during October, with the maximum count to date of 118 occurring in

FIGURE 8. AVERAGE COMBINED MONTHLY MAXIMA OF GADWALL, 1980-95.

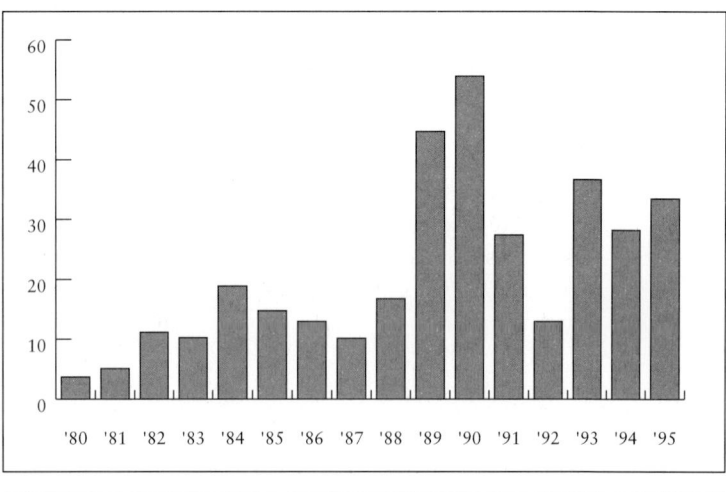

FIGURE 9.
AVERAGE
MAXIMA OF
GADWALL BY
MONTH
(1980-95).

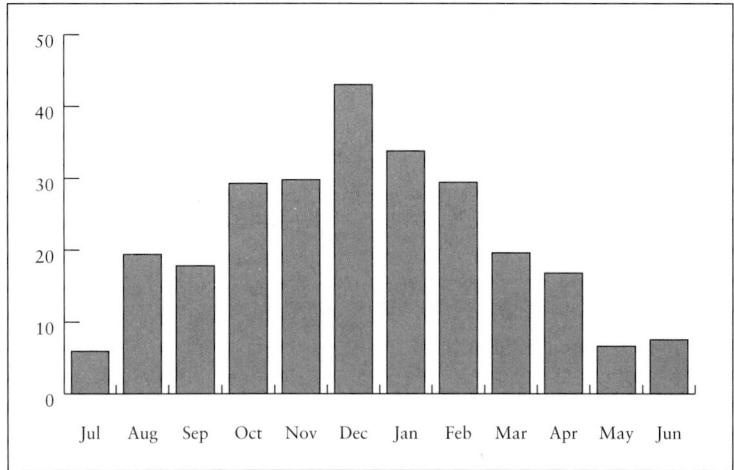

FIGURE 9. AVERAGE MAXIMA OF GADWALL BY MONTH (1980-95).

October 1990 at a time when water levels were relatively low. During the last five years there has been a tendency for the largest populations to occur in the early part of the new year, rather than December as in previous times.

A general increase is in line with the population trend of the species in southern England. Birds seem to be driven away by cold weather (in marked contrast to Wigeon) and this could mask any trends suggested by the records.

Teal *Anas crecca* (A)

Common winter visitor, formerly bred.

Historically this species was present throughout the year with small numbers breeding. Recently, occurrences at the reservoirs in May and June have become decidedly unusual and breeding was last proved during the fieldwork for the *Hertfordshire Breeding Bird Atlas* (Mead & Smith 1982).

There is some evidence that over-stocking with domestically reared Mallards adversely effects the breeding fortunes of the species. When this practice has ceased in the past Teal have, under favourable circumstances, returned to breed (Sage 1959).

The first returning birds of the autumn arrive from late June with numbers building up over the following months. Between November and January the size of the visible population largely depends on water levels. Low water during the winter 1989/90 resulted in a count of over 660, the largest to date. A few birds remain until the end of April.

This species generally prefers the sewage farm, Wilstone and Tringford Reservoirs but can also be numerous at Marsworth Reservoir when water levels are low.

Mallard *Anas platyrhynchos*

Abundant breeding resident.

Several thousand Mallard-type ducks are released annually during the late summer for 'sporting' purposes. These birds totally shroud any movements of wild birds. Pairs nest at all the reservoirs.

Pintail *Anas acuta*

Scarce winter visitor and passage migrant.

Pintails have been recorded in all months between August and April with peaks in November and January/February. Occasional birds occur outside this period.

The majority of sightings are of one to five birds but flocks of up to 16 occur, mainly during February. The largest flock to date is of 57 birds departing from Wilstone Reservoir to the south-west on 31 January 1985 (Lack & Ferguson 1993). On a number of occasions intensive study of the large Mallard flocks for the monthly wildfowl count has resulted in the discovery of a previously undetected female or immature Pintail. Birds seems to prefer Tringford and Wilstone Reservoirs, these waters attracting the largest numbers of the commoner species of dabbling duck.

Garganey *Anas querquedula*

Scarce visitor.

Although Garganeys are almost annual visitors they remain somewhat enigmatic to many bird watchers and are always a good bird to find. Records have generally occurred from mid-March to mid-June and from the end of July to October with slightly more birds seen during the latter period. The largest count to date is of 11 birds in August 1977.

Exceptionally Garganeys have occurred in the breeding season, although there has been no more than circumstantial evidence of actual breeding in recent years. The only documented breeding at Tring Reservoirs took place in 1928 (Sage 1959).

Very unusual winter records refers to individuals from 5 January until 12 February 1990 and another from 21 October 1993 until 8 January 1994.

*Blue-winged Teal *Anas discors*

Rare vagrant, one record.

1978:   a male, 5 April at Wilstone Res

The only record was the (first or) second in Hertfordshire considered to be of wild origin. There has been only one subsequent record in the county and there are no Buckinghamshire records.

Shoveler *Anas clypeata* (A)

Common visitor, rare in summer.

Shovelers now breed only rarely and in small numbers at Tring Reservoirs (for example a brood of five young was seen in 1988) although counts during May and June have varied from one to as many as 25 birds. As with Teal, pressure from the release of large numbers of domestically reared wildfowl seems to reduce the chances of successful breeding by Shovelers.

In parallel with the occurrence of other ducks, numbers of Shovelers increase during August and generally peak during the following two months. In winter, between November and the new year, over 50 birds generally remain at the reservoirs before the early spring return passage between late January and March. The maximum count in recent times was of 300 in February 1986. In the past, Shovelers showed a distinct preference for Marsworth Reservoir but increased disturbance in recent years appears to have made the site less attractive to the species.

FIGURE 10. AVERAGE COMBINED MONTHLY MAXIMA OF SHOVELER, 1980-95.

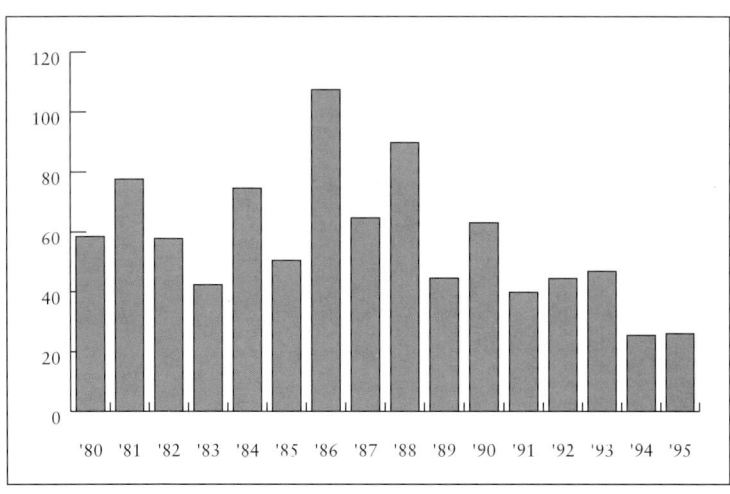

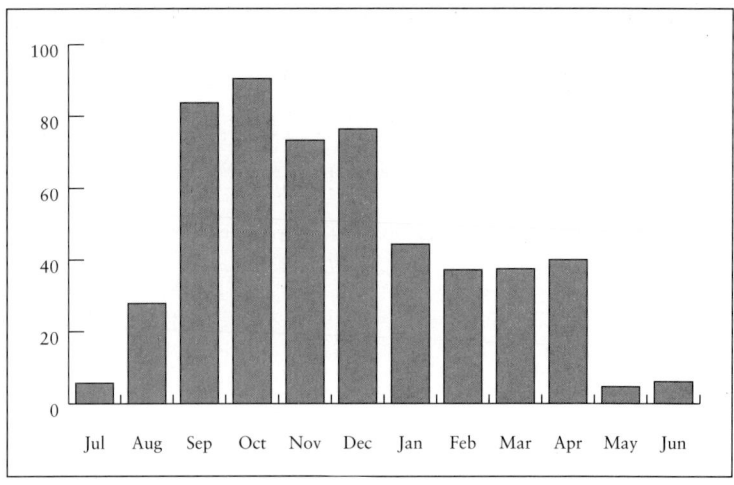

FIGURE 11.
AVERAGE
MAXIMA OF
SHOVELER BY
MONTH
(1980-95).

Red-crested Pochard *Netta rufina*

*Scarce
visitor of
unknown
origin*

There are several records of single birds remaining for extended periods, although two birds were noted in October 1994. One of the recent records, a male, was said to have arrived as a juvenile during 1988. This bird spent the majority of its time amongst the Mallards and was noted as having mated with a number of female Mallards over several years. Despite disappearing for days, or even weeks at a time, it was reported at least monthly until the last sighting on 6 October 1991. Whilst most records will refer to naturalised birds or recent escapes from captivity, it is possible that wild birds have reached Tring Reservoirs. A flock of 'about 40' is said to have been present in the autumn of 1889 or 1890 (Hartert & Jourdain 1920).

A hybrid Red-crested Pochard × Pochard was identified during October 1989 and remained until the following February.

Pochard *Aythya ferina* (A)

*Common
resident
breeding in
small
numbers.*

Pochards have bred at Tring Reservoirs since 1850, the first recorded nesting in Hertfordshire. However, they now nest only in small numbers, usually less than five pairs, and breeding success is often very poor. After the breeding season numbers begin to build in August/September and peak at over 200, usually between October and January with high counts of 600 in October 1971 and 692 in

November 1995. By April most of the winter visitors have departed. Birds generally prefer Wilstone and Startops End Reservoirs in winter but the few breeding pairs show a preference for the quieter parts of Wilstone and Tringford Reservoirs.

*Ring-necked Duck *Aythya collaris*

Rare vagrant, one record.

1977:   male, 2-30 April, intermittently at Wilstone and Marsworth Res

It is now considered that only a single individual was involved in the 1977 sightings, contra Gladwin & Sage (1986). 1977 was the first year in which over five individuals of this duck were reported in Britain and marked the beginning of a large increase in records. It is therefore a little surprising that there have been no further records at this site.

There are four other Hertfordshire and five Buckinghamshire records of this North American duck.

*Ferruginous Duck *Aythya nyroca*

Rare vagrant, six records.

1965:	male, 3-4 January
1971:	female, 7-9 November
1972:	female, 18 March
1972:	two males and a female, 25 June -2 July
1993:	male, 16-21 October, Wilstone Res
1995:	female, 30 October-10 January 1996, Wilstone Res, intermittently at Startops End Res and College Lake

Some doubt exists with all records of this species as to the origins of these birds although some, such as the 1993 and 1995 individuals, arrived with groups of Pochard at appropriate times of year for wild birds to be considered.

There are 14 other Hertfordshire and four Buckinghamshire records.

Tufted Duck *Aythya fuligula*

Common breeding resident.

Tufted Ducks have bred at Tring Reservoirs since the first recorded occurrence of nesting in Hertfordshire during 1893. In total there are usually 50-100 birds present during May and June and between one and 10 pairs breed annually with young being raised on all waters, although recently relatively few broods have fledged. Hatching appears to be a few weeks later than Pochard.

**MARSWORTH
RESERVOIR IN
SUMMER.**

Autumn movement begins relatively early, with moulting flocks often present between July and September when up to 308 have been recorded. The wintering flocks peak during December and January with 300 or more birds usually present with a recent maximum of 578 in December 1985. There is a noticeable return passage during March and April.

*Scaup *Aythya marila*

Rare migrant and winter visitor.

Small numbers of Scaup occur in most years, with the majority of records falling between October and April. Parties, generally of one to four birds, occur amongst the gatherings of other *Aythya* ducks. A 'large flock' was present during December 1900 or January 1901 and there is a record of 16 on 31 March 1954 at Startops End Reservoir. The highest count in later years was of eight birds (three males and five females) on Marsworth Reservoir on 16 April 1974. Some birds remain until the spring or until hard weather forces further movement, others are merely day visitors.

In July 1912, a female was seen with two ducklings. This bird had been wounded earlier in the year and may have mated with a Tufted Duck (Hartert & Jourdain 1920).

Observers should be aware of hybrids resembling this species and that female Tufted Ducks can show extensive white face patches.

Aythya hybrid *Aythya sp.*

Status uncertain.

This family of diving ducks seems particularly prone to hybridisation. Some astonishingly confusing or misleading individuals are produced which show either a combination of characteristics from different species or closely resembling one parent or another, often rarer, species.

All unusual looking birds, or individuals appearing to show characteristics of rarer species, require careful study for correct identification. Detailed field notes should be taken to assist in record adjudication.

Hybrids have been proven to be more common than previously supposed and may be scarce winter visitors. For example, during the winter 1990/91 at least five different individuals were identified. They varied in appearance from obvious hybrids to birds very like Scaup and Ferruginous Duck.

*Eider *Somateria mollissima*

Rare vagrant, two records.

| 1963: | male, 3 May |
| 1993: | male, 31 October, Tringford Res was part of a large influx to inland waters |

There are five additional Hertfordshire and three Buckinghamshire records. The latter records relate to two birds at nearby Weston Turville Reservoir in May 1988 and birds also involved in the 1993 influx.

*Long-tailed Duck *Clangula hyemalis*

Rare winter visitor.

With 14 records prior to 1980 it is surprising that the following are the only subsequent reports:

1980:	one first noted on 31 December 1979 remained until 19 January, Wilstone Res
1988:	23 October until 4 June 1989 – an astonishingly long staying bird – remained mainly at Startops End Res
1993:	19-23 October, Wilstone Res

By comparison there have been 10 records elsewhere in Hertfordshire and nine in Buckinghamshire since 1973. The earlier Tring Reservoirs records show a distinct peak from late October to late November.

Common Scoter *Melanitta nigra*

Scarce, mainly spring migrant.

This species occurs almost annually with about 80 per cent of the records falling in April and May, often during foggy weather. Since 1980 reports have been spread throughout the year with the exception of February. The largest groups have been 15 on 23 April 1989 and 26 on 29 May 1992.

Birds tend to favour the larger waters of Startops End and Wilstone Reservoirs.

*Velvet Scoter *Melanitta fusca*

Rare vagrant, three records.

1930:	two females, 3-4 November, Wilstone Res
1963:	male, 8-17 December
1991:	female, 23-24 October, Startops End Res

The 1991 bird appeared during a significant influx of this species to inland waters in south-east England.

The four Buckinghamshire and five additional Hertfordshire records (of seven birds) have all been during the period October to January.

Goldeneye *Bucephala clangula*

Common winter visitor.

Goldeneyes generally return to the deeper reservoirs towards the end of October. Numbers build to around 20 between January and March, with birds often present into April, and odd ones remaining for another month. Most of these late birds are young individuals. The maximum number counted recently was 25 during January 1985 during a spell of hard weather.

Exceptionally a male was present from mid-July to mid-September 1933, an immature male remained through the summer of 1990 and an injured female was resident between 1993 and 1996.

Smew *Mergus albellus*

Rare winter visitor.

Formerly a more regular winter visitor, this smart duck now occurs only sporadically with sightings largely governed by hard weather.

Birds have been recorded in 11 years since 1980. The highest recent counts are of seven birds at Wilstone Reservoir during a prolonged spell of freezing conditions in February 1991 and up to three birds, including an adult male, in January/February 1996. The vast majority of records have been in the first two months of the year although they have been noted between the beginning of December and early March.

*Red-breasted Merganser *Mergus serrator*

Rare winter visitor.

This largely maritime species occurs during the winter months, in groups of up to three birds. The majority of recent records have been on the deeper reservoirs with eight records since 1980 in the period between late October and late March.

Observers should be aware of the similarities between females of this and the following species.

Goosander *Mergus merganser*

Winter visitor.

Goosanders occur annually, if sporadically, between October and March, usually involving between one and five birds. These can sometimes remain on the deeper waters for several weeks, and individuals may linger until May. Exceptionally an injured female remained until at least mid-summer in 1996. In past years much larger flocks have occurred, for example 71 in February 1960 and 63 in January 1970 but with a recent peak of only 24 in January 1985.

Ruddy Duck *Oxyura jamaicensis*

Resident species breeding in small numbers.

This accidentally introduced species initially bred at Tring Reservoirs from 1965-68 and more regularly from 1975, with non-breeding birds peaking at 77 in the winter of 1983/4. Since then smaller numbers have been present throughout the year, especially during winter, and the reservoirs have been little affected by the subsequent increase of this species in Hertfordshire (Figure 12).

Each year a few pairs breed. There appears to be passage during both migration periods with recent maxima of 24 in September 1991 and over 30 in April 1986.

FIGURE 12. WINTERING POPULATION OF RUDDY DUCKS AT TRING RESERVOIRS AND IN HERTFORD-SHIRE (PER HNHS 1993).

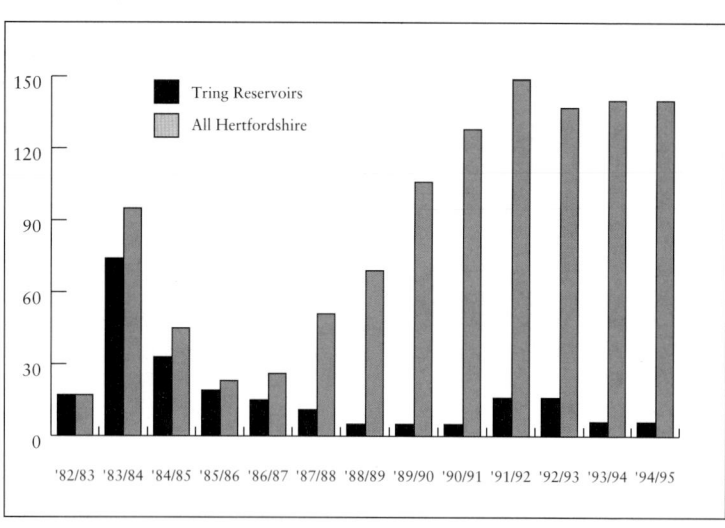

*Honey Buzzard *Pernis apivorus*

Rare vagrant, one record.

1996:   21 May, Wilstone Res

This was the first record for Tring Reservoirs. There have been 14 other records in Hertfordshire and 10 in Buckinghamshire.

*Red Kite *Milvus milvus*

Rare vagrant, three records.

This species was undoubtedly more common in the distant past but by the mid-19th century it had become very rare in the region (Lack & Ferguson 1993).

The existing records probably relate to wandering continental birds but in the future these will perhaps be supplemented by birds from the English re-introduction schemes. Released birds and some of their offspring have a coloured plastic tag attached to the carpal joint of the wing and observers should note any details seen.

Late 1860s: shot near Wilstone Res
1981:   20 April, flying over north-west
1987:   15 March, flew south over Marsworth Res

*Marsh Harrier *Circus æruginosus*

Scarce, mainly spring migrant.

The first Hertfordshire record was of a male shot at Wilstone Reservoir on 21 May 1935 with a further record in 1958. Since 1965 this species has become a regular spring visitor at the reservoirs with birds sometimes roosting in the Wilstone reedbed overnight. These individuals can be seen quartering the reeds during the early morning but will frequently depart after feeding. Since 1980 there have been

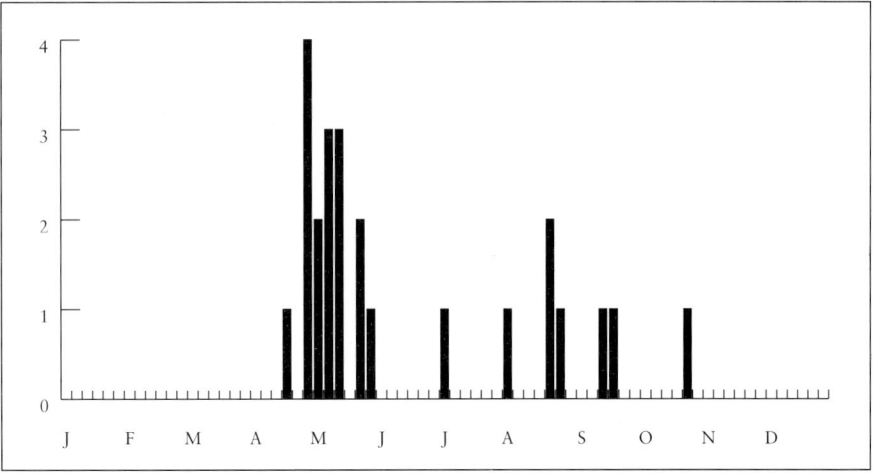

FIGURE 13. THE OCCURRENCE OF MARSH HARRIER BY FIVE DAY PERIODS (1980-95).

records of birds in all months between the extreme dates of 20 April and 27 October. However, half the records have occurred in the first four weeks of this period.

*Hen Harrier *Circus cyaneus*

Rare vagrant, twelve records.

There are six records between 1884 and 1980 with a further six occurring since. Recent records have all been between mid-October and the end of March with the exception of a bird on 11 May 1991. Hen Harriers generally pass straight through but a male remained in the area for several days during severe winter weather in early 1985.

*Montagu's Harrier *Circus pygargus*

Rare vagrant, two records.

1977:   26 May, female flying north-east
1978:   13 May

Both records have been in spring involving birds that flew straight through or lingered only briefly. This species appears to be increasing as a breeding bird in Britain and more frequent occurrences may be hoped for.

Sparrowhawk *Accipiter nisus*

Resident breeding species.

The relative frequency of sightings, and increasing population of this species today, belies the severe pressure under which it has been in the past. Previously the subject of persecution from keepers, the population crashed during the 1960s due to the effect of organochlorine pesticides.

With at least three pairs nesting in the area, birds can now be seen regularly around the woodland at each of the waters. Sightings are most reliable when males perform their display flights in February and March. Birds are also attracted to the winter roosts of various species with the Starling gathering at Wilstone Reservoir a favourite spot.

Buzzard *Buteo buteo*

Rare migrant.

This scarce migrant has been reported more frequently in recent years, reflecting the increase in the British population and occasional nesting in the region. Records have occurred in six different months since 1980, with birds appearing mainly between late August and December with additional March and May records.

*Osprey *Pandion haliætus*

Rare, mainly spring migrant.

This species has been recorded almost annually in recent years, although generally these are migrants that pass quickly through the area. Records occur mainly between late April and the end of May and seem to be increasing in line with national trends. Ospreys are much scarcer during autumn migration with only four records in August and September.

One colour-ringed bird that remained at the reservoirs (mainly at Tringford Reservoir) for at least six days in May 1992 had hatched at Loch Garten, Speyside, two years previously. Birds that do linger are probably individuals too young to breed and generally occur later in the spring than birds of breeding age.

Kestrel *Falco tinnunculus* (A)

Breeding resident.

A few pairs breed every year, either within the reservoirs area or nearby, and can be seen during any season hunting in the surrounding farmland.

During late 1994, an all dark (melanistic) bird was reported from the farmland around Wilstone Reservoir.

*Merlin *Falco columbarius*

Rare migrant, has wintered.

A typical record of this species involves a bird dashing low over one of the reservoirs and disappearing over a bank or hedge never to be seen again. Although not annual, there have been 12 records since 1980, mainly in winter during the period 19 October-8 March. A bird which arrived 10 November 1985 subsequently over-wintered.

There has been one spring record on 25 April 1978 and an early autumn record of a female or juvenile on 19 August 1990.

Hobby *Falco subbuteo*

Increasing summer visitor and migrant.

Birds at Wilstone Reservoir often give excellent views making this site one of the best in the region for watching this species. The first birds arrive back in the area in the last two weeks of April with the spring passage lasting through to the first half of June. Hobbies can often be seen at close quarters around the Wilstone hide from early May. The first juvenile birds start to appear around the middle of August with records through until early October.

As many as 12 birds can be seen hawking for insects in the spring with smaller numbers in autumn when birds take the occasional hirundine. In line with the national trend, numbers have increased markedly in recent years.

No breeding has been proven within the confines of the reservoirs area but is known to occur nearby and this presumably accounts for the records of birds throughout the mid-summer period.

*Peregrine *Falco peregrinus*

*Rare
visitor.*   This species was formerly an occasional autumn or winter visitor, with 19 records noted by Hayward (1947), but had not been recorded since October 1952. There have been two recent records and hopefully these may increase as the national population continues to grow.

1988:   immature female, 19 February, Wilstone Res
1990:   18 & 24 September, Wilstone Res

The first record relates to a bird believed to have been in the area since

December 1987. With any record of a large falcon, caution should be taken to eliminate escaped falconers' birds which could include hybrids with other species.

Red-legged Partridge *Alectoris rufa*

Scarce and decreasing resident.

Red-legged Partridges are seen sporadically in the farmland surrounding the reservoirs. Records, mainly of one to five birds, are most numerous in April and September although this bias is probably accounted for by the relative ease in seeing territorial birds, new families and the low height of the crops during these months. A family party in 1991 suggested local breeding but a flock of 24 in October 1994 may have resulted from the introduction of birds for shooting.

Grey Partridge *Perdix perdix* (R)

Rare and decreasing species.

Previously described as 'quite often seen on adjacent fields' (Holdsworth *et al* 1978), the Grey Partridge has now become a distinctly rare bird and may indeed no longer be resident in the area. With fewer than 15 sightings since 1980, the majority of birds have been seen in the months between April and July. The largest recent flock however was an exceptional count of 15 in November 1992. An adult with a juvenile in July 1989 was the latest occurrence suggesting breeding, although a male was heard calling nearby in July 1994.

Quail *Coturnix coturnix* (R)

Rare visitor, two recent records.

1986:   19-24 August, Wilstone Res

1994:   male, 14 May, near Wilstone Res

Prior to 1980 the previous records were in July 1965 (Holdsworth *et al* 1978). Strangely neither recent record occurred during 'good' Quail years and notably no birds were recorded in the area during 1989, when up to four pairs bred at Aldbury, a few kilometres east of Tring Reservoirs.

Pheasant *Phasianus colchicus*

Common breeding resident.

This species is common in the farmland surrounding the reservoirs, with the population being artificially maintained through the feeding and release of captive-bred birds.

*Lady Amherst's Pheasant *Chrysolophus amherstiæ*

Rare visitor of feral or escaped origin, one record.

1984:   male, 1 April, Wilstone Res

Although birds no longer seem to be present at Mentmore, Buckinghamshire and Whipsnade, Bedfordshire, and the main population is considered to be sedentary, males have been known to disperse perhaps over several kilometres (Cramp & Simmons 1980) which could account for this record. Birds were present at Mentmore, which is only six kilometres from Wilstone Reservoir, up to 1985 (Lack & Ferguson 1993) and there have been other recent records in the area from Ashridge Forest and Wiggington, Hertfordshire.

Water Rail *Rallus aquaticus* (A)

Resident, breeding in small numbers.

This species is best seen when water levels are sufficiently low for there to be mud bordering the reedbeds but at other times it can be very difficult to observe. Birds can often be heard calling ('sharming') from the reedbeds, especially early in the morning. Due to its secretive nature, it is difficult to assess the true population level but 11 birds were counted at Marsworth Reservoir on 10 January 1989.

There is little doubt that the species breeds at Wilstone and Marsworth Reservoirs most years and this has been confirmed by the appearance of juvenile birds near the hide at Wilstone.

A SPOTTED CRAKE AT TRINGFORD RESERVOIR, APRIL 1996.

*Spotted Crake *Porzana porzana*

Rare vagrant, seven records.

1883:	September, shot
1885:	October, shot
1949:	2 April, Wilstone Res
1973:	Mid August-November
1977:	3-20 November, Wilstone Res
1990:	15 September, Marsworth Res
1996:	12-25 April, Tringford Res

The elusive nature of this species could no doubt have masked considerably more birds than have been recorded, although they apparently have always been uncommon here.

[*Little Crake *Porzana parva*

Possibly a rare vagrant.

1887:   5 January – one said to have been shot at the reservoirs

The skin has been reported as lost for many decades. The record was considered by Sage (1959) to be 'at the least very doubtful'.]

*Corncrake *Crex crex*

No recent records.

The Corncrake almost certainly bred in the area of the reservoirs in the past with, for example, records in the months May to July for the years 1910-1918. Sporadic records continued until September 1944 when the last recorded bird was shot when mistaken for a rabbit in the undergrowth.

Unless agricultural practices change greatly it is unlikely that this nationally declining species will ever again give its rasping call from fields by the reservoirs.

Moorhen *Gallinula chloropus*

Common breeding resident.

True numbers of the species are masked by its relatively elusive nature and its preference for the cover provided by the reedbeds and the vegetation at Tringford Reservoirs in particular. Systematic wildfowl counts however have found in excess of 100 birds at Tringford during hard weather or low water levels and there seems little doubt that large numbers are present throughout the year.

The size of the breeding population is similarly hard to evaluate but newly hatched chicks can be seen at various sites every year. As with other local sites, the population is presumably supplemented to an unknown extent by influxes of migrant and wintering birds.

Coot *Fulica atra*

Abundant breeding resident.

Unlike most other waterfowl, Coot numbers tend to reach their peak in the period between September and November. The average flock size during the last quarter of the year has reduced considerably over the past decade with the counts in excess of 1000 birds, as witnessed at

FIGURE 14. AVERAGE COMBINED MONTHLY MAXIMA OF COOT, 1980-95.

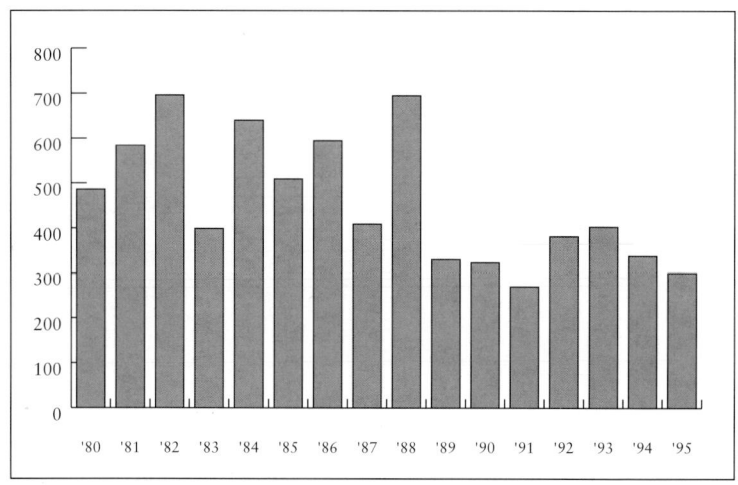

FIGURE 15. AVERAGE MAXIMA OF COOT BY MONTH (1980-95).

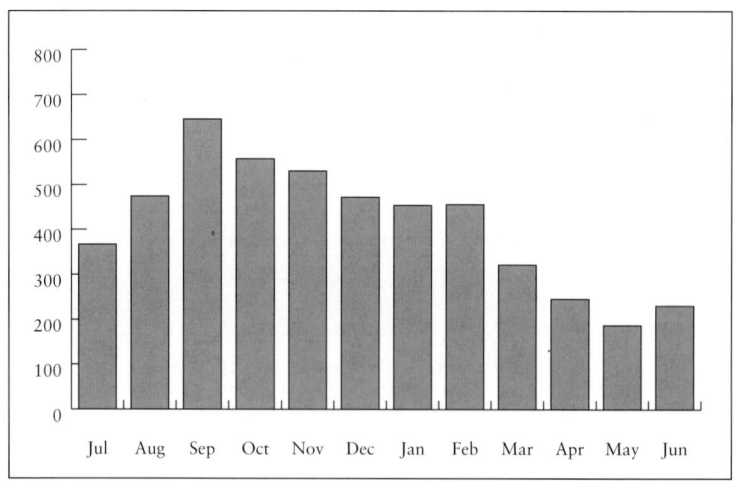

times in the early to mid 1980s, now unlikely. Previous studies suggested mid-winter to be the period of greatest abundance except when large areas were frozen (Hayward 1968).

During the breeding season at least 150 birds remain in the area although many of these must be non-breeding birds as numbers of nesting pairs are not thought to exceed 40.

*Crane *Grus grus*

Rare vagrant.

1996:   adult, 20 May, Wilstone Res, flew north-east

This is the first record for Hertfordshire and there are three Buckinghamshire records.

Oystercatcher *Hæmatopus ostralegus*

Scarce migrant and winter visitor.

Oystercatchers have occurred in most recent years and there are noticeable peaks in records at migration times in April/May and August. Birds can, however, occur through the winter. Normally only present in ones or twos, the largest group recorded was of 15 birds on 5 September 1949. There has been a growth in the number of records recently which could reflect the increasing tendency of this species to be seen inland but has, no doubt, also been affected by the increase in the number of observers.

*Black-winged Stilt *Himantopus himantopus*

Rare vagrant, one record.

1987:   pair, 16 May, sewage farm

It seems probable that this record, involving a pair seen mating, were the celebrated birds that appeared at Holme, Norfolk the next day and remained to breed, raising two young.

*Avocet *Recurvirostra avosetta*

Rare migrant, six records.

1973:	28 June
1974:	two, 28 April
1984:	24 March, Startops End Res
1984:	12, 2 December, Wilstone Res
1993:	5 May, Wilstone Res
1996:	12 April, Startops End Res

The record in 1973 was the first for Hertfordshire although, as the national population increases, presumably occurrences at Tring Reservoirs should also. It is possible that the 12 birds seen flying over Wilstone Reservoir in December 1984 were on their way to join the increasing winter population in the south-west.

Even high water levels need not deter this species and the bird in May 1993 spent five hours swimming at Wilstone Reservoir.

Little Ringed Plover *Charadrius dubius*

Scarce migrant.

As mentioned in the introduction, the first recorded breeding of Little Ringed Plover in Britain occurred at Startops End Reservoir in 1938, when a pair successfully reared three young. However, more typical habitat for this species is provided by mineral extraction in Hertfordshire and Buckinghamshire river valleys and the species is now generally a scarce migrant at Tring Reservoirs. In the right conditions birds can stay to breed and three pairs nested at Wilstone Reservoir in 1976. Little Ringed Plovers arrive early, in March, but the greatest numbers occur during the autumn when in good conditions as many as 20 have been seen together in September. When water levels remain high, however, birds do not visit and the species is not recorded in some years. The latest date for a bird in recent times was one on 10 September 1989.

LOWER WATER LEVELS ARE LIKELY TO ATTRACT LITTLE RINGED PLOVERS TO THE RESERVOIRS IN AUTUMN.

Ringed Plover *Charadrius hiaticula*

Scarce visitor. Spring passage can commence as early as late February with numbers building up to a peak in May, although it is on a very small scale. Odd birds (which could be late migrants or failed breeders) can occur in June and significant autumn migration starts in July, with the main passage in August and September. A flock of 41 was seen (in August 1943) but the largest recent count has been of 33 birds in August 1989. Numbers quickly reduce in October although occasionally birds can turn up during the winter months. Ringed Plovers are seen at Tring Reservoirs in most years but there have been blanks, such as 1981 and 1986.

There is a record in November 1904 at Marsworth Reservoir of the subspecies *C. h. tundræ* from northern Scandinavia and Siberia (Witherby 1934).

Kentish Plover *Charadrius alexandrinus*

Rare vagrant, two records.

| 1964: | 16 August, Startops End Res |
| 1976: | one found late on 22 April with two present from 23-24th, reducing to one on the 25 April |

These are two of only three Hertfordshire records and there is a single Buckinghamshire record from 1981.

Golden Plover *Pluvialis apricaria*

Variable winter visitor and spring migrant. Golden Plovers can often be found associating with flocks of Lapwings in fields adjacent to the reservoirs during the winter months, although numbers are somewhat variable due to the availability of suitable feeding conditions. Small numbers start to appear in September and can build up to quite sizeable flocks by November and December, when counts of 400 are regular. The largest recent count was of 639 birds in December 1993. Numbers are maintained between January and March and flocks can still number over 300 in April, but then they quickly reduce. When water levels allow, a few birds often join the bathing and resting Lapwings on the causeway in front of the hide at Wilstone Reservoir.

Passage birds, sometimes in complete summer plumage, can occur until May, long after the wintering population has moved on.

Grey Plover *Pluvialis squatarola*

Rare visitor.

Grey Plovers put in irregular appearances at Tring Reservoirs. Most have occurred from March to May and August to October, with the highest numbers in September. The largest party was of four birds on 23 September 1990 at Marsworth Reservoir. There have been two recent winter records, both of two birds, on 17 December 1991 and 3 January 1993.

*Sociable Plover *Chettusia gregaria*

Rare vagrant, one record.

1961:   29 October, Wilstone Res

The only record for Hertfordshire, this winter-plumaged bird stayed for a mere 10 minutes before leaving with a flock of Lapwings (Devlin et al 1962).

Lapwing *Vanellus vanellus*

Abundant winter visitor and migrant.

Lapwings frequent the fields and reservoir margins and can often be seen flying over the area. The local wintering population can be quite high in the early part of the year, when flocks up to 3000-4000 have been seen. In March and April numbers begin to fall with only a few pairs remaining to breed. In the recent past these have included breeding at the sewage farm where up to five pairs have produced young in years when suitable habitat is present.

Autumn movements always start early for Lapwings with noticeable numbers in June which continue to build up into August. Larger flocks occur again towards the end of the year. Cold weather often produces sizeable movements when substantial numbers can pass through the area.

Knot *Calidris canutus*

Rare migrant.

The 12 occurrences prior to 1980 involved eight in autumn, between August and November, and two records in both winter and spring. There have been seven recent records since 1980 with two spring and four autumn sightings involving up to 14 individuals. The autumn birds occurred between mid-August and mid-September. There was an exceptional flock of 45 over Wilstone Reservoir on 7 September 1978.

Sanderling *Calidris alba*

Rare migrant.

Sanderlings mainly occur at Tring Reservoirs in spring with all recent records between the middle and the end of May. There has been one autumn bird, in October 1986, and a single winter record of a bird present at the sewage farm on 31 December 1968.

The maximum number recorded was a flock of seven birds on 13 May 1976.

Little Stint *Calidris minuta*

Rare migrant.

Little Stints have occurred in seven years since 1980, mostly in autumn between July and late September with a peak in September. The largest group in that period was six birds together in September 1981, coinciding with an influx into the London area. The largest flock recorded, however, was of 18 birds on 18 September 1953. There was a late individual on 4-5 November 1990 and there have been two spring records of single birds on 27 April 1966 and 6 May 1967.

Temminck's Stint *Calidris temminckii*

Rare vagrant, seven records.

1939:	24 May
1943:	28-29 August
1949:	25 September
1950:	24 October-3 November
1952:	two, 14-15 September
1976:	6 June
1977:	23-24 May

The great increase in observers over the last 10 years has not improved the frequency of reports of this genuinely rare migrant with no records since the spring of 1977. This is in contrast with Buckinghamshire where one site has produced 10 spring records since 1980. This can undoubtedly be attributed to the lack of suitable muddy conditions at the reservoirs in spring.

*Pectoral Sandpiper *Calidris melanotos*

Rare vagrant, six records.

1949:	14 September
1969:	19 October-13 December
1973:	3 September-10 October

1986:	11 October, sewage farm
1988:	juvenile, 29-30 September, Wilstone Res
1989:	adult female, 9-19 September, Startops End and Wilstone Res

The commonest Nearctic wader, records of Pectoral Sandpiper at Tring Reservoirs closely follow the national picture, with arrivals falling within a narrow band between early September and mid-October.

The bird present in September 1989 was ringed and incomplete views of the ring number, coupled with moult characteristics, suggested that this was 'almost certainly' the adult female ringed at Marston Sewage Farm, Lincolnshire, on 29 August 1989.

Curlew Sandpiper *Calidris ferruginea*

Rare migrant.

Almost exclusively autumn migrants at Tring Reservoirs, Curlew Sandpipers have occurred in seven years since 1980. Up to seven were present at the sewage farm in September 1969 during an exceptionally heavy autumn passage in Britain. Apart from two adults in August 1989, all recent records have been of juveniles in September and October. There are two historic spring records in early May.

Purple Sandpiper *Calidris maritima*

Rare vagrant, two records.

| 1959: | 25 October |
| 1968: | 2 September |

This species is a very irregular visitor to inland areas. There is one other Hertfordshire and two Buckinghamshire records, all between August and November.

Dunlin *Calidris alpina*

Irregular migrant and winter visitor.

Probably the commonest small wader to be seen at Tring Reservoirs, birds have occurred in most years with particularly good numbers between 1989 and 1991. However, the fact that they usually occur in small flocks boosts the annual totals and there may be no birds present for many months, particularly in years with high water levels. There is a marked peak in spring numbers during April, but only a single record in June (in 1989). The protracted autumn passage is from July through to November, with a few birds appearing during the winter months.

The largest flock of recent times was of 33 in November 1991, but about 100 were seen in November 1970.

*Broad-billed Sandpiper *Limicola falcinellus*

Rare vagrant, one record.

The only record was the first for Hertfordshire, on a distinctly unusual date for this species in Britain:

1946:   6 October, Wilstone Res

A subsequent Hertfordshire record at Rye Meads and Broxbourne in July and August 1958 (Gladwin & Sage 1986) does not appear to have been submitted to the *British Birds* Rarities Committee and cannot therefore be substantiated.

Ruff *Philomachus pugnax*

Scarce migrant and winter visitor.

A highly variable wader in size and colour, spring birds pass through between late April and mid-May with a spring high of nine birds on 26 April 1991. Autumn passage occurs from early July until October and birds can remain for extended periods when good feeding conditions prevail. There is a total of five winter records between December and February including an overwintering bird in 1990/91 which fed on grain put out for wildfowl and gamebirds at Tringford Reservoir. In some years there are no records but in 1989, when conditions were particularly favourable, birds were present for almost a month with up to 13 present on 27 August. The highest number on record was 30 at Marsworth Reservoir on 3 October 1976 (Gladwin & Sage 1986).

Jack Snipe *Lymnocryptes minimus* (A)

Rare autumn and winter visitor.

The Jack Snipe is a secretive and difficult species to observe and is best looked for when water levels drop a little below the edge of the reedbeds; Marsworth Reservoir appears to be a fairly regular site under these conditions. Previously recorded as a regular winter visitor (Holdsworth *et al* 1978), numbers are obviously difficult to assess, but it seems likely that up to four birds have wintered between early October until February. The maximum count on record was of six at Marsworth Reservoir on 19 January 1969.

This species may be declining at Tring Reservoirs; one from November 1995 was the first since 1992.

Snipe *Gallinago gallinago*

Common winter visitor and migrant.

Although cryptically camouflaged, Snipe are usually fairly obvious when water levels are low. They also occur in the damp marginal vegetation around the reservoirs, along the small streams and even in open, grassy fields where they are much more difficult to see unless they take to the air. These factors make it hard to assess the overall population, and although commonly seen they are probably not particularly numerous. However, numbers can fluctuate in adverse weather conditions, either concentrating birds at open feeding areas or forcing them to leave the area completely when ice closes over. The highest recorded count is of 200 at Wilstone Reservoir in March 1949 and no recent counts have come close to this level.

Breeding was suspected in 1923 and 1943 when drumming was heard but was not proven until 1944 when at least one pair nested at Wilstone Reservoir (Sage 1959). Single birds occasionally oversummer.

Three albinos have been seen at the reservoirs.

*Great Snipe *Gallinago media*

Rare vagrant, one record.

1941:   5 November, Wilstone Res

There are nine records for Hertfordshire, the last in 1952, and a single Buckinghamshire record in 1962.

*Long-billed Dowitcher *Limnodromus scolopaceus*

Rare vagrant, one record.

1977:   juvenile, 22 October-12 November, Wilstone Res

The first and only record for Hertfordshire, this bird was perhaps the same individual seen at Staines Moor, Middlesex, on 1-15 October.

Woodcock *Scolopax rusticola* (A)

Scarce, winter visitor.

Occasional birds are seen in winter. Woodcocks are probably under-recorded because of their secretive behaviour and the lack of public access to most of the woodland areas around Wilstone and Tringford Reservoirs. Up to 12 were reported at Wilstone Reservoir on 16 February 1993.

Breeding was suspected in 1986.

Black-tailed Godwit *Limosa limosa*

Scarce migrant.

Black-tailed Godwits have been seen in 10 of the years since 1980. Parties of up to eight birds have occurred from early-July to early November with a peak in July. There are only two recent spring records and thus a single flock of 15 in March 1985 was unusual. When conditions are suitable, birds may be tempted to remain for a few days.

Bar-tailed Godwit *Limosa lapponica*

Scarce migrant.

This species is slightly scarcer at Tring Reservoirs than the Black-tailed Godwit. With both species it appears that fewer but larger flocks occur in spring, including a single flock of 27 Bar-tailed Godwits flying over in May 1983. This species shows a very concentrated spring peak during the first week of May. During the autumn they occur more frequently but in smaller parties. The autumn peak is in September and there are recent winter records of single birds in February 1985 and January 1987 during hard weather.

Whimbrel *Numenius phæopus*

Frequent migrant.

Whimbrels occur during both migration seasons but are generally only seen (or heard at night) flying over. Peak spring passage, of small flocks with up to five birds, falls between mid-April and late May. In the autumn the greatest numbers are seen in August although passage can span the period between July and early-September. At this time of year larger numbers are recorded such as a flock of 52 flying over in August 1982, the largest Hertfordshire flock to date.

Curlew *Numenius arquata*

Frequent migrant and winter visitor.

Curlews have been seen in every calendar month, but only an average of about seven birds have been seen in the years since 1980. There are marked peaks in numbers in March/April and August but a few Curlews also appear during the winter months. June and July are usually blank months, with birds recorded in only four years – the eight in June 1991 being the only record for that month. No doubt these were migrants, but breeding does occur in the Vale of Aylesbury, Buckinghamshire (Lack & Ferguson 1993). The largest recorded flock at Tring Reservoirs was of 25 birds passing south-west in August 1969. As with Whimbrels, many of the records are of birds flying over.

Spotted Redshank *Tringa erythropus*

*Rare
migrant.*

In recent years a Spotted Redshank has only appeared once in spring, an individual seen twice in April 1989. Most of the records are between August and mid-September with a single bird in October 1984. There is an unusual record of a wintering individual which remained from September until 21 December 1969. Spotted Redshanks have been seen in seven years since 1980 with the majority being records of single birds though there were three in August 1985. The maximum count is an exceptional 10 in August 1976.

Redshank *Tringa totanus*

*Frequent
visitor.*

Even when this species is not visible, it can makes its presence known with its yelping call. Redshanks have not occurred during the early

64

months in recent years, but migrants appear in March and continue to pass through in April and May.

Records in June may refer to migrants but Redshanks bred at Tring Reservoirs in 1909, 1935-6, 1938, 1944 and 1946 (when there were four pairs) and still do so nearby at College Lake and Pitstone No 2 Quarry when conditions are suitable.

Autumn passage is light but prolonged, with birds present through to the end of the year, tailing off in December, although they may remain throughout the winter if conditions are suitable. Generally spring birds outnumber autumn ones by at least 50 per cent.

There have been annual records in recent years with the largest counts of six in March 1986 and November 1991.

*Marsh Sandpiper *Tringa stagnatilis*

Rare vagrant, one record.

1887:   October

The first record for Britain, this individual suffered the usual fate of rare birds at the time and was shot. Sadly the specimen was destroyed in 1890. The only other record in Hertfordshire is of a single bird at Broxbourne Gravel Pits in April 1984.

Greenshank *Tringa nebularia*

Frequent migrant.

Like the Redshank, the first indication of the Greenshank's presence is often its distinctive call. A classic migrant, records of Greenshanks fall into two obvious groups in spring and autumn, with autumn birds outnumbering those in spring by over four to one. Peak passage is in April/May and August/September, with numbers quickly falling off during late September and October. The largest numbers recorded are three in spring at the sewage farm in April 1984, and in autumn a maximum of 12 birds during August 1989. There have been records for all years since 1980, with higher numbers in years with good feeding conditions such as 1989 and 1990.

*Lesser Yellowlegs *Tringa flavipes*

Rare vagrant, one record.

1953:   18-23 September, Wilstone Res

This individual occurred during a minor 'invasion' of this North American species.

*Solitary Sandpiper *Tringa solitaria*

Rare vagrant, one record.

1984:   juvenile, 5 and 12 October, Marsworth Res

The report of this elusive individual forms the second record for Hertfordshire, the first being at Rye Meads in 1967. A subsequent bird was seen at Kimpton Mill in August 1989, making Hertfordshire a surprisingly popular venue for this American species.

Green Sandpiper *Tringa ochropus*

Frequent migrant.

Green Sandpipers frequent the muddy margins and concrete banks of the reservoirs and can be relatively numerous in years with lower water levels and therefore good feeding conditions. Occurring in all months, there are definite peaks in numbers during migration times in April and August, with birds present from mid-June onwards presumably being early returning migrants. One bird was suspected of over-summering in 1990 – an unprecedented event.

If conditions are mild and water levels suitable, small numbers of Green Sandpipers occasionally overwinter, with up to six birds present in January 1989. The highest number recorded was 24 in September 1974 and recently 13 in August 1989.

In recent years the sewage farm has provided the most reliable mud and has therefore been the site most frequented by this species during the autumn.

Wood Sandpiper *Tringa glareola*

Rare migrant.

Wood Sandpipers occur rarely between July and September, with a marked peak of records in August. Only in 1981 have spring records outnumbered autumn when two passed through together in May. The highest number on record is of nine on 22 August 1943. Strangely no Wood Sandpipers appeared in the good wader year of 1989.

Common Sandpiper *Actitis hypoleucos*

Frequent migrant.

Common Sandpipers have been recorded in every month in recent years. Spring passage starts in earnest in April and continues through until May when up to 14 birds have occurred in loosely associated parties. July sees the return of the first birds from the breeding grounds.

The largest numbers are found in the autumn with a broad peak from July to September; records quickly tail off in October. The highest count is of 37 birds in September 1974 and recently 12 on two dates in August 1989.

Odd birds have occurred in winter; one was seen from 16 January to 4 February 1991 and another was present in December 1994.

Turnstone *Arenaria interpres*

Rare migrant.

Small numbers of birds have appeared since 1980 with 10 spring records in eight years, mainly between the end of April and the end of May.

Autumn records in the same period were limited to a single in

August 1990 and five records in August 1989 including a flock of eight, the second largest flock recorded in Hertfordshire.

Turnstones appear to be attracted down to rest even when water levels are unfavourable. The party of eight at Wilstone Reservoir in August 1989 were found clinging to the near-vertical embankment.

*Red-necked Phalarope *Phalaropus lobatus*

Rare vagrant, four records.

1885:	October
1948:	female, 2-7 June
1966:	25 August
1982:	female, 2-9 May, Startops End and Wilstone Res

There are two other Hertfordshire and three Buckinghamshire records. All records fall between May and October.

*Grey Phalarope *Phalaropus fulicarius*

Rare vagrant, 10 records.

Most Grey Phalaropes have occurred in the months of October and November with records also in August and late September. There have been the following records since 1961:

1982:	11-12 November, after gales
1987:	two, 16-17 October, Wilstone Res, one remaining until the 19th, following the 'Great Storm'
1989:	a moulting adult, 25 August, Wilstone Res
1994:	juvenile, 31 October, Wilstone Res

Unlike the Red-necked Phalarope, this species appears during late autumn passage after gales. The 1989 record was unusually early but arrived during a tremendous rainstorm and was present for only an hour or so.

*Pomarine Skua *Stercorarius pomarinus*

Rare vagrant, one record.

| 1928: | adult, 22 November, flew south-west over 'Little Tring' |

It is not certain if the place in question was Tringford Reservoir, also known as Little Tring, or the hamlet of that name a short distance away. This late date is a typical for this species.

There has been one other Hertfordshire and four Buckinghamshire records.

*Arctic Skua *Stercorarius parasiticus*

Rare vagrant, two records.

1946:   12 May, Wilstone Res

1985:   23 August, Startops End Res

The relative scarcity of this species inland is somewhat surprising given the fact that this is the commonest skua on the east coast.

There are 11 other Hertfordshire and two Buckinghamshire records.

A JUVENILE LONG-TAILED SKUA, FOUND DEAD ON 16 SEPTEMBER 1988.

*Long-tailed Skua *Stercorarius longicaudus*

Rare vagrant, three records.

1919:	juvenile, 27-30 August
1985:	juvenile, 1-3 September, Wilstone Res
1988:	juvenile, 14-15 September, Wilstone Res, later found dead

There is a tendency for a few juveniles of this species to occur inland in autumn. Birds often linger for several days, perhaps because they are in a weak physical condition.

There is one other Hertfordshire and two Buckinghamshire records.

*Great Skua *Stercorarius skua*

Rare vagrant, three records.

1962:	24 October
1983:	3 September, Wilstone Res
1987:	16 October, Wilstone Res, after the 'Great Storm'

The occurrence of this species seems to be particularly related to the incidence of gales off the south and west coasts of Britain.

There are two other Hertfordshire records and one Buckinghamshire record.

Skua sp. *Stercorarius sp.*

There has been much discussion regarding the identity of six birds which passed over Wilstone Reservoir 17 August 1990. These, when added to the fact that the 1988 bird was originally thought to have been an Arctic Skua and only re-identified when found dead, amply illustrate the difficulty in the specific identification of inland skuas. Full and detailed field notes should be taken of all such birds.

Recent studies at the Wash have shown that numbers of skuas migrate overland along the Nene and Great Ouse river systems in autumn (Easy 1994). There is very little evidence of this passage at inland sites.

*Mediterranean Gull *Larus melanocephalus*

Scarce migrant and winter visitor.

In line with the marked increase in both national and regional records (Young 1991), there has recently been a marked upsurge in the number of records at the reservoirs, although this can probably be attributed in part to the great increase in observer awareness and knowledge of gull identification over the past few years.

THERE IS A
CHANCE OF
FINDING A
MEDITER-
RANEAN GULL
AMONGST THE
BLACK-HEADED
GULLS COMING
IN TO ROOST
AT WILSTONE
RESERVOIR.

Following the first record in 1983 there have been at least 16 individuals noted with four or five in the 1994/95 winter. All the recent records have been in the Wilstone Reservoir gull roost but the earliest recorded autumn date related to a juvenile bird at the sewage farm on 3 September 1988. All roost records have been in the period 18 October to 9 April and have comprised mainly of first-winter birds with the exception of single birds in second-winter and second-summer plumages and two adults.

Little Gull *Larus minutus*

Migrant occurring in variable numbers.

Much more frequent since the 1950s, this diminutive gull now appears annually. Always more common in spring, up to 35 birds have been recorded in a day during the period between the end of March and mid-June. The most concentrated passage is generally in late April with the largest flock of 30 on 24 April 1993.

71

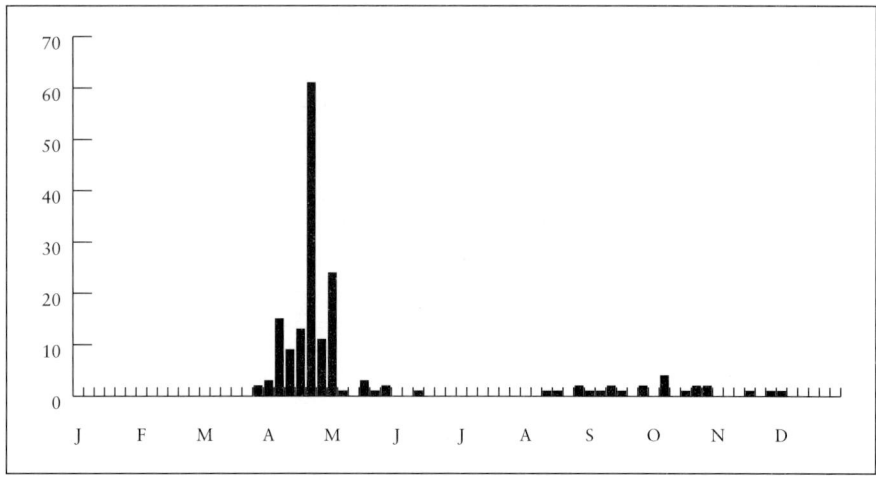

FIGURE 16. THE OCCURRENCE OF LITTLE GULL BY FIVE DAY PERIODS (1980-95).

Autumn passage is very sporadic and usually involves small numbers of adults and juveniles in the period from mid-August to the end of October.

Occasional wandering birds have occurred in all months.

*Sabine's Gull *Larus sabini*

Rare vagrant, four records. This usually pelagic bird occurred at Tring Reservoirs, as well as a number of other inland sites, after the 'Great Storm' of 1987 when large numbers were displaced from the Bay of Biscay. Remarkably there was another report two years later:

1987:   two adults, 16 October, Marsworth Res

1987:   first-winter bird, 17 October, in the gull roost

1987:   two different adults, 18-21 October, roosting at Wilstone Res with one remaining until 23 October, feeding on flooded fields during the day

1989:   an adult, 22 November, passed eastwards over Startops End Res

There are three other Hertfordshire and two Buckinghamshire records relating to birds displaced in October 1987 and an additional bird at Hilfield Park Reservoir, Hertfordshire in September 1988. All records are pre-dated by two birds at Willen Lake, Buckinghamshire on 13 October 1981.

Black-headed Gull *Larus ridibundus*

Abundant visitor with small numbers present during the summer.

This species makes up the majority of the gull roost which has been present at Wilstone Reservoir in the autumn and winter months since the 1930s. Although a few birds occur throughout the summer months, the roost starts to form in September with a large increase the following month. A sudden return influx occurs at the end of June, but there is no early autumn roost of moulting birds as at other roost sites (Elton 1992-6).

The winter peak of 20,000 or more birds occurs between November and January. The 1993 BTO gull roost survey counted 20,232 gulls, all but a few hundred being of this species. Subsequent counts have estimated the roost to be several times this figure on exceptional days. Although few of the wintering birds remain, by the end of March there is a noticeable spring passage with peak estimates from the roost of 25,000 in March 1978 and 14,000 in March 1991.

Common Gull *Larus canus* (A)

Abundant winter visitor and migrant.

Common Gulls are much less abundant in the gull roost than the previous species and, while a few individuals turn up with the first returning Black-headed Gulls, there is no appreciable build-up until late October. Significant numbers appear in November with generally up to 1000 at the Wilstone Reservoir roost during December and January although numbers can vary greatly from day to day. Up to 4000 were present in January 1994 and there was an estimate of 7000 in January 1995. March usually sees an increase in roosting birds with passage also visible during the day. Relatively few are seen in April and the species is very scarce in the summer months.

Lesser Black-backed Gull *Larus fuscus*

Common migrant, and winter visitor in small numbers.

Few 'large' gulls occur at Tring Reservoirs with this species being by far the most numerous. Peak numbers occur during August to November and January to May. Autumn maxima are generally in the range 110 to 170, but with 300 in October 1977 and August 1981; late winter/spring numbers are no more than half that number.

There have been few records of races apart from the British *L. f. grællsii*, with odd reports of birds resembling one of the Scandinavian races *L. f. intermedius/fuscus*, presumably the former which has the more westerly migration route.

Herring Gull *Larus argentatus*

Scarce winter visitor and migrant.

Herring Gulls are surprisingly uncommon at the reservoirs. Small numbers, usually less than five, join the winter roost on occasions and birds are sometimes observed passing over. In recent years no flock of more than 28 birds has been reported although there is a record of 139 in February 1979. As most birds occurring are immature, it is possible that the species has been overlooked in the past. With advances in the identification of immature gulls, it is hoped that this will be less problematic in future.

As with Lesser Black-backed Gulls, relatively few attempts have been made to fully analyse the racial mix of birds although it is known that numbers of the large Scandinavian race *L. a. argentatus* occur in winter. A few records relate to birds resembling the Yellow-legged Gull *L. (a.) michahellis*, although it is debatable if birds at roost are safely identifiable without clear views of leg colour, particularly from December when all Herring Gulls start to acquire cleaner head patterns. Yellow-legged Gulls should be looked for in late summer and early autumn when numbers peak in south-east England.

*Glaucous Gull *Larus hyperboreus*

Rare vagrant, two records.

1942:   6 March, Wilstone Res
1977:   31 December

The low number of 'large' gulls using the reservoirs is reflected by the rarity of this species and the absence of confirmed Iceland Gull records. This is in contrast to the Marston Vale clay pits, Bedfordshire, less than 25 kilometres away, which is one of the most reliable winter sites in the Midlands for 'white-winged' gulls.

Great Black-backed Gull *Larus marinus*

Scarce winter visitor and migrant.

A rare vagrant prior to the mid 1940s, this species remains a scarce visitor. Odd birds or small parties have been recorded passing over in all months except June and September, or have joined the winter roost. The highest recent count is of four birds in January 1995. This is in contrast to the numbers relatively close by in the Marston Vale, Bedfordshire, where there were, for instance, 300-400 at Brogborough Lake in January 1994 (Brind 1995).

Kittiwake *Rissa tridactyla*

Occasional spring migrant and rare visitor.

This graceful pelagic gull is now recorded annually in small numbers suggesting a regular cross-country migration. The majority have been seen during the period March to May but there are records from all months except October and November, although very few have been seen in the latter half of the year.

Occurrences are usually of single birds passing quickly through but on occasions birds join the roost or remain at the site for a few days. One bird roosted at Wilstone Reservoir between 2-27 March 1990 and spent the daytime feeding in a nearby field. Parties of 40 birds on 28 April 1985 and 24 at Wilstone Reservoir on 15 March 1992 were part of movements of this species across south-east England.

Sandwich Tern *Sterna sandvicensis*

Rare migrant.

Birds have been recorded in nine years since 1980. As with most terns, sightings are more frequent in spring, mainly during April. However, larger parties are sometimes reported during the autumn, particularly in September. The largest counts to date have been at least 16 on 22 September 1990 and 13 on 24 September 1972.

*Roseate Tern *Sterna dougallii*

Rare vagrant, one record.

1968:   18 May

This is the only Hertfordshire record, but this species has been recorded in Buckinghamshire on four occasions.

Common Tern *Sterna hirundo*

Common migrant and common summer visitor.

The first records of this species are usually in the second week in April and birds are then reported throughout the following six weeks. Numbers vary considerably, with some years producing large flocks of up to 80 birds (on 1 May 1991) and others no more than 10 or so.

A few birds, clearly breeding locally, visit during the summer months and occurrences have increased greatly in recent years with the successful establishment of the breeding colony at College Lake. In 1994, the provision of a tern nesting platform at Wilstone Reservoir resulted in the first successful breeding of this species at Tring Reservoirs.

Return passage commences as early as the end of July and continues over the following month or so. Numbers tend to be considerably lower than in the spring with rarely more than 20 birds present.

Arctic Tern *Sterna paradisæa*

Migrant in variable numbers.

Arctic Terns generally arrive a week to 10 days after the first Common Terns and peak in late April. Numbers are variable but can include impressively large flocks of over 100 birds on rare occasions, with a peak of 115 at Wilstone Reservoir on 3 May 1991. Recent studies have shown that Arctic Terns are most likely to appear inland when faced with head winds either locally or for a large part of their prior migration route through the English Channel (Kramer 1995).

Autumn migration usually involves just a few birds between August and October, with a peak of 20 on 17 August 1985. There is a tendency for records to occur later than Common Tern at this season.

Birds prefer Startops End and Wilstone Reservoirs.

Little Tern *Sterna albifrons*

Scarce migrant.

Single birds have been recorded in about half of the years since 1980 but their appearances are usually brief and unpredictable. Spring passage occurs from mid-April to early June with occasional autumn records in August and September. Up to three birds can occur in any year but 1991 exceptionally produced six birds between 12 April and 8 June.

Conditions which result in Little Tern records are less clearly defined than with the other scarce terns with some appearing in settled weather.

Black Tern *Chlidonias niger*

Variable but frequent migrant.

The Black Tern is not as numerous now as 30 years ago when Tring Reservoirs used to see large flocks, for example 125 on 21 May 1948 and 100 or more on 2 May 1958 and 31 May 1966.

Birds first arrive in the last two weeks of April, although the largest number of records relate to periods of south-east winds during May. By the end of that month migration has tailed off, after which there are a few June and July records. In spring 1991 observers recorded 74 bird/days between 16 April and 3 June. The highest recent spring count is of 38 birds on the relatively early date of 23 April 1994.

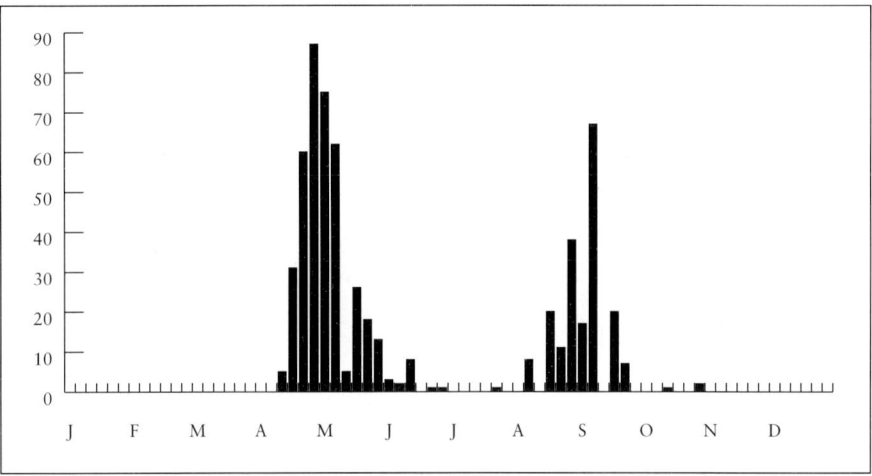

FIGURE 17. THE OCCURRENCE OF BLACK TERN MEASURED AS BIRD DAYS BY FIVE DAY PERIODS (1980-95).

Records from the end of July and early August probably refer to the autumn passage which peaks in late August and mid-September, with stragglers appearing until the middle of October. The latest recent date is 1 November 1992. Although rarely in large numbers at this time of year, there has been exceptional flocks of 65 birds on 11 September 1992 and 90 on 15 September 1994.

*White-winged Black Tern *Chlidonias leucopterus*

Rare vagrant, three records.

1929:	an adult and an immature bird, 7 October
1994:	juvenile, 24-27 August, Wilstone Res
1994:	juvenile, 15 September, Wilstone Res

There is one other record of this eastern species in Hertfordshire.

*Little Auk *Alle alle*

Rare vagrant, two records.

1841:	December, shot
1965:	4 November, Wilstone Res

In addition a bird at Wilstone Reservoir on 2 February 1986 arrived by car and stayed a few hours. It had been rescued from nearby Wingrave, Buckinghamshire.

*Puffin *Fratercula arctica*

Rare vagrant, one record.

1961:   18 September

There have been about 30 records in Hertfordshire and Buckinghamshire usually involving storm driven individuals.

Feral Pigeon *Columba livia*

Abundant resident.

Substantial flocks of Feral Pigeons, involving up to 400 individuals, are generally resident in the vicinity of the duck feeding sites where they capitalise on the availability of easy food.

Stock Dove *Columba œnas* (A)

Common breeding resident.

Stock Doves are numerous in the area with birds present around much of the farmland and especially in the woodland to the south of Tringford Reservoir. They breed in holes in trees and can be found in the breeding season wherever suitable habitat is present. Although no quantitative information is available, there seems every reason to believe that the species is increasing in the area in line with regional trends (Smith *et al* 1993).

Hard weather movements can bring additional birds to the reservoirs with flocks of 111 during December 1990 and 137 the following December.

Woodpigeon *Columba palumbus*

Abundant breeding resident.

Woodpigeons are always numerous in the area around the reservoirs. During the summer large numbers breed in suitable habitat – average densities suggest that at least 20 pairs nest in the immediate vicinity. In winter flocks of several hundred birds are quite frequently seen, the largest recent gathering being of 1500 in November 1992.

During the autumn, large movements can occur. These often coincide with other large-scale movements across the region.

Collared Dove *Streptopelia decaocto*

Common resident.

This species is only a relatively recent addition to the local avifauna having first arrived and bred in 1961. In the past, concentrations of Collared Doves were attracted to spilt grain at New Mill, but with improvements in efficiency this no longer seems to be the case. The largest flocks now gather around James Farm, adjacent to the roundabout near Startops End Reservoir. The numbers present here vary but up to 50 birds are usual in the winter months, with an exceptional record of more than 300 present in January 1995.

Turtle Dove *Streptopelia turtur* (R)

Scarce migrant. The Turtle Dove was formerly far more common than it is today, with 5-10 pairs said to have been breeding during the late 1970s (Holdsworth *et al* 1978) and three pairs still breeding at the sewage farm in the early 1980s. The decreasing numbers reported from the reservoirs closely mirror the local and national trends. In recent years there have been only a few spring records but 1993 was the first year in recent times in which no birds were reported.

Exceptionally there is a winter record of one at the sewage farm on 17 January 1985.

Ring-necked Parakeet *Psittacula krameri*

Rare visitor, five records.

1977:	22 October
1979:	13 October
1991:	6 October, Tringford Res
1992:	11 January, Wilstone Res
1993:	17 February, Wilstone Res

The stronghold in Britain of this garrulous alien is south of the River Thames though there has been a slow increase in the number of records in Buckinghamshire and Hertfordshire in recent years. It can be expected to occur with greater frequency at the reservoirs in the future.

Cuckoo *Cuculus canorus*

Common summer visitor. A traditional harbinger of spring, Cuckoos are often first recorded at the reservoirs during mid-April. Birds are then obvious until they stop singing in July and juveniles are occasionally recorded during August. The area around Marsworth Reservoir is often a good place to see this species. Birds are attracted to the colony of Reed Warblers in the reedbed and use the trees along the bank as song and display posts. The breeding population is hard to assess but it is believed that there are around four singing males in the area most springs.

Barn Owl *Tyto alba* (A)

Rare visitor. The status of this species at Tring Reservoirs has reflected the well-documented national decline. However, there has recently been a slight increase in the number of occurrences, with one or two birds present in the area in the spring and summer of recent years. It is possible that this apparent increase is due in part to re-introduction schemes.

This species generally hunts over open fields but has also been recorded over the reedbed at Wilstone Reservoir.

Little Owl *Athene noctua*

Resident breeding species. Little Owls are the most visible of the owls at Tring Reservoirs and can often be seen sitting out during the day at favourite sites. Birds are particularly obvious during March and April before the trees come into full leaf and at this time individuals can generally be found at traditional sites, particularly around Wilstone Reservoir.

In recent years this species appears to have bred annually, with up to five pairs on territory in 1994. There has been an increase in records

LITTLE OWLS CAN FREQUENTLY BE SEEN DURING THE DAY AT WILSTONE RESERVOIR.

over the last few years which is probably related to both observer awareness and an actual rise in the population.

Birds prefer mature trees with large holes and often favour ivy-clad pollarded trees in hedgerows, but also require the close proximity of open ground in which to hunt. The farmland around Wilstone Reservoir offers several such combinations of habitat.

Tawny Owl *Strix aluco*

Resident breeding species.

This species nests annually in woodland around the reservoirs and occasionally fledged juveniles can be seen roosting in the trees to the south of Wilstone Reservoir. Tawny Owls are generally nocturnal and are most often heard at night although visitors to the reservoirs after dark are rare and the species is probably under-recorded. However, birds can sometimes be found perched at dusk or seen flying across roads in the area during the night.

The species may have decreased in recent years as the population of no more than three pairs is less than previously reported (Holdsworth *et al* 1978). Birds can be found at both Tringford and Wilstone Reservoirs and are at their most vocal and obvious during the first couple of months of the year when territories are being established and pair formation takes place.

*Long-eared Owl *Asio otus*

Rare vagrant, two records.

1885:   August
1886:   December

The lack of suitable habitat around the reservoirs means there is unlikely to be a change in this species' status as a rarity here although summering birds have been present nearby in both Hertfordshire (Young pers obs) and Bedfordshire (Dazley & Trodd 1994).

*Short-eared Owl *Asio flammeus*

Rare migrant and winter visitor.

Short-eared Owl records occur mainly between the start of November and the end of April although two of the 15 records have been in August and one in September. While the earlier records might relate to birds wintering nearby and straying close to the reservoirs, there can be little doubt that more recent records involve passage birds or individuals dispersing from their breeding areas.

81

**TRINGFORD
RESERVOIR IN
WINTER.**

*Nightjar *Caprimulgus europæus*

Rare vagrant, two records.

1985:   male, 5 August, Wilstone Res

1986:   5 June, Wendover Arm

The rare status of this species is no surprise given the lack of suitable habitat around the area.

Swift *Apus apus*

Common non-breeding summer visitor.

Although a common summer visitor, no Swifts breed at the reservoirs. Large flocks are seen in spring, mainly in May, when deteriorating weather conditions often bring birds to the reservoirs to feed. Estimates of up to 5000 birds have been made although large concentrations of this species are notoriously difficult to count accurately.

Swifts usually arrive in the second half of April and are recorded until late September, although most have departed by the beginning of that month. The latest record relates to a single bird at Wilstone Reservoir on 1 October 1982, almost one month after the main population had departed.

The numbers of Swifts ringed at the reservoirs is dependent on suitable weather conditions prevailing. The blustery unsettled May of 1994 resulted in over 130 birds being ringed whereas the more clement May 1995 resulted in only 11 being caught. However, two of these were birds previously ringed at Startops End Reservoir in the spring of 1993 and 1994, indicating a degree of site fidelity and the importance of the reservoirs as a food source. One bird ringed at Startops End Reservoir during May 1994 was found dead 12 days later after an untimely collision with a grave stone. During the intervening period it had moved 63 kilometres to Odiham, Hampshire.

Kingfisher *Alcedo atthis* (A)

Resident breeding species.

Kingfishers are regularly present around the reservoirs but, even though brightly coloured, can be difficult to see and are often first picked out by their piercing flight call. They occur throughout the year at all of the reservoirs, although Tringford and Wilstone are particularly favoured. The population appears to be fairly stable with at least two pairs breeding in the vicinity. During September and October family parties can often be watched at Tringford Reservoir.

There is some evidence of movement with records of ringed birds moving to or from Northampton and Surrey. At one stage a bird trapped at the reservoirs held the accolade of being the longest recorded Kingfisher movement, being recovered 250 kilometres away from the original ringing site at Knaresborough, North Yorkshire.

*Hoopoe *Upupa epops*

Rare vagrant, one record.

1950:   30 April, Wilstone Res

While there is only this single record at the reservoirs, a bird at Wilstone Great Farm on 22 October 1991 was almost within viewing distance from the reservoir bank.

*Wryneck *Jynx torquilla*

Rare vagrant, six recent records.

1975:	10 September, Wilstone Res
1977:	September
1981:	6-12 September
1982:	3 October
1987:	10 August
1989:	9 October, Marsworth Res

In line with its local status, the Wryneck can now be classed as an extremely rare visitor during autumn migration. There seems little doubt that, during the last century, Wrynecks must have occurred more frequently when the species was classed as a 'common summer visitor' in the region (Kennedy 1868), although there is no further documentary evidence to support this.

The few records fit well within the regional pattern of occurrence.

Green Woodpecker *Picus viridis* (A)

Scarce, non-breeding, resident.

Numbers of this species tend to be highest outside the breeding season and particularly in early autumn. At this time of year it is likely that there is a dispersal of juveniles from their breeding areas in the large tracts of woodland nearby.

Although representing no more than circumstantial evidence of local breeding, a juvenile Green Woodpecker was seen at Wilstone Reservoir during the summer of 1995 and a pair held territory the following year.

Great Spotted Woodpecker *Dendrocopos major*

Resident breeding species.

During late winter and early spring a number of drumming birds can be found marking out territory and it is believed that at least four pairs breed. Formerly classed as 'scarce' (Holdsworth *et al* 1978), the population has increased in line with the national and local trend (Smith *et al* 1993; Lack & Ferguson 1993). The main areas for nesting are the woodland to the south of Tringford Reservoir and in the private area south of Wilstone Reservoir. The bird's undulating flight and sharp 'tchik' call are a common feature in all parts of the area.

Lesser Spotted Woodpecker *Dendrocopos minor*

Former resident now scarce visitor.

Holdsworth *et al* (1978) reported 2-4 pairs present with breeding recorded at Marsworth, Tringford and Wilstone Reservoirs. Since that time numbers have greatly declined and birds are now seen only rarely. This may be due in part to the disappearance of mature diseased Elm trees from areas such as the woodland around Tringford Reservoir. Single birds were recorded in 11 of the years between 1980 and 1995 but the last record of more than one bird together was of three in July 1982. Most reports refer to late winter or early spring, when drumming and calling males are most conspicuous.

A bird over-wintered at Tringford Reservoir during 1992/93 and there followed a series of records of drumming males both at that site and at Wilstone Reservoir the following spring. It is not clear if one or more birds were present in the area but the capture of a juvenile at Wilstone Reservoir during July 1993 raises the possibility that both the presence of a female and subsequent breeding went undetected.

Skylark *Alauda arvensis* (R)

Common resident and passage migrant.

Skylarks breed in the fields immediately surrounding the reservoirs, with up to 10 singing males present in recent years. There is, however, little past information to compare this with and populations may be reducing in line with the national trend.

Birds are also recorded in small flocks during passage periods and much larger movements can be associated with hard weather during the winter; for example flocks in excess of 100 birds were noted on 9 December 1990 and 10 February 1991. A flock of a similar size on 10 October 1992 presumably consisted of birds on migration.

Sand Martin *Riparia riparia*

Common passage migrant.

Often the first migrant to be reported in the spring, Sand Martins usually return during the second week of March. The bulk of spring passage occurs during April and May and, after a serious decline in numbers during the past two decades, the reservoirs are now enjoying some large gatherings of this species. The largest recent spring flocks recorded have been of 1000 birds on 12 May 1989 and 2200 on 12 May 1990.

Recent numbers on autumn passage have also been encouraging with sizeable flocks occurring between mid-August and early

September. When bad weather forces birds down to the reservoirs to feed, movements have included gatherings of 700 birds (2 September 1989 and 15 August 1990) and a large flock estimated to contain 1300 birds on 6 September 1992. Migrants pass through until the middle of October with the latest bird in early November.

Swallow *Hirundo rustica* (A)

Common summer visitor and passage migrant.

Swallows usually arrive during the first two weeks of April although there have been a few records in late March. The bulk of passage occurs during the latter half of April and continues through May. Recent high counts include 800 on 22 April 1989, and 1500 on 12 May 1990.

During the summer, a few pairs of Swallows can be found breeding in the farm buildings adjacent to Wilstone and Startops End Reservoirs.

Large numbers also occur during autumn although for unknown reasons there has not been a roost of hirundines at the reservoirs since 1982. Passage often continues into mid-October but there is an exceptionally late record of a juvenile bird at Marsworth Reservoir on 11 and 14 November 1991.

A Swallow ringed at Marsworth Reservoir was found dead at Nador, Morocco, a distance of 1876 kilometres from Tring Reservoirs.

*Red-rumped Swallow *Hirundo daurica*

Rare vagrant, one record.

1981:　　17 May, Wilstone Res

The four other Hertfordshire records include one at Aldbury near Tring Reservoirs on 11 June 1949.

House Martin *Delichon urbica*

Common passage migrant and summer visitor.

The first migrants usually return during early April, although there are a few records in the last week of March. Numbers build to a peak in early May with a maximum movement of 4000 birds on 12 May 1990.

A small colony exists at the Lock Keeper's Cottage, Marsworth, with 12 nests in 1995.

Autumn passage is often evident during September with movements abruptly reducing in the middle of October.

Tree Pipit *Anthus trivialis*

Rare
passage
migrant.

1989:	three, 8 April, Wilstone Res
1989:	15 April
1993:	one, on territory during May and June, Grand Junction

With only three recent records, this species is surprisingly rare at the reservoirs considering that breeding areas exist as close as Ashridge Forest and Wendover Woods.

Meadow Pipit *Anthus pratensis*

Mainly a
frequent
passage
migrant.

Spring passage occurs from early March until mid-April. Autumn passage is most apparent between mid-September and late October, although the actual numbers migrating through the area during both passage periods are variable and difficult to quantify.

There are two records of singing birds being present at Marsworth Reservoir and the sewage farm during the summers of 1970 and 1978 respectively and, more recently, in 1993 a male held territory on farmland adjacent to the Grand Junction. Meadow Pipits used to breed on the canal banks prior to 1920 but there has been no evidence of successful breeding since.

Small numbers winter in the area, particularly near the sewage farm and Tringford Reservoir.

Rock Pipit *Anthus petrosus*

Scarce,
mainly
autumn
migrant.

There is an historic record of a Rock Pipit at Startops End and Marsworth Reservoirs on 16 October 1932 (Hayward 1947). They have been recorded almost annually since 1980 mainly during the period from late September to early November, with most records occurring during October and a daily maximum of five birds on 12 October 1990.

In spring, a few birds have been recorded between late March and May. A bird on 29 March 1981 is the only one identified as being of the Scandinavian race *A. p. littoralis* although it is possible that most birds occurring at Tring Reservoirs are in fact of this subspecies.

This narrow band of passage is in line with records from other local sites such as Willen Lake, Buckinghamshire (Lack & Ferguson 1993), Hilfield Park Reservoir, Hertfordshire (Elton 1992-6), and Farmoor Reservoir, Oxfordshire.

Water Pipit *Anthus spinoletta*

Rare migrant and winter visitor.

As with Rock Pipit, this species is recorded during October, although records tend to occur from the middle of the month. Water Pipits have been recorded less than annually since 1980 with birds occurring between mid-October and the beginning of April with a daily maximum of three on 27 December 1990. Occasionally birds have stayed for protracted periods such as one during the early part of 1991 when water levels were low. This pattern of occurrence during autumn and winter is similar to that at other local sites in Buckinghamshire (Lack & Ferguson 1993).

When confronted with any potential Rock or Water Pipit, observers should beware of the difficulties involved in separating the Scandinavian race of Rock Pipit *A. p. littoralis* from Water Pipits in spring and from the nominate race of Rock Pipit *A. p. petrosus* in autumn.

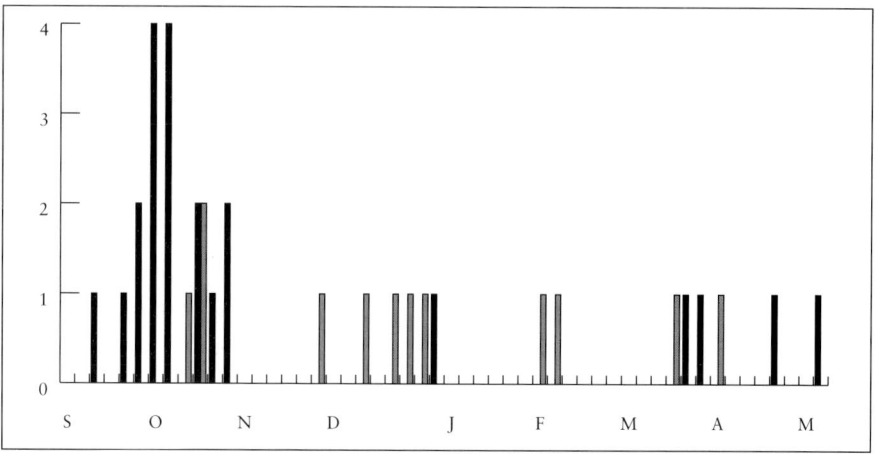

FIGURE 19. THE OCCURRENCE OF ROCK (BLACK BARS) AND WATER PIPITS (GREY BARS) MEASURED AS ARRIVALS BY FIVE DAY PERIODS (1980-95).

Yellow Wagtail *Motacilla flava*

Common passage migrant and summer visitor.

Spring passage of Yellow Wagtails commences from early April and continues into early May with birds being found on the open grassy areas particularly around Wilstone and Startops End Reservoirs. The earliest Hertfordshire record is of one at Wilstone Reservoir on 23 March in 1991. Peak passage is during the second half of April, with flocks of over 20 almost annual since the early 1980s. The

YELLOW
WAGTAILS ARE
A FREQUENT
SPRING
MIGRANT AT
STARTOPS END
RESERVOIR.

maximum count to date is of 105 on 22 April 1988.

In recent years local breeding has occasionally been suspected with adults seen carrying food for young in mid-summer.

Autumn passage probably commences from late July and is usually heavier than that recorded in spring, with flocks of up to 50 birds occurring. Examples include 45 birds on 17 September 1985 and 50 on 20 September 1990. Passage lasts until early October.

Birds showing the characteristics of the Blue-headed Wagtail M. f. flava are recorded occasionally in spring. A male was seen repeatedly collecting and flying off carrying food on 15 June 1967.

Birds apparently showing the characteristics of Syke's Wagtail M. f. beema have been recorded four times in the spring. It is possible that these individuals are variants of M. f. flava or hybrids between flava and flavissima.

Grey Wagtail Motacilla cinerea

Breeding resident, passage migrant and winter visitor. During the early part of the century, the Grey Wagtail was considered to frequent the reservoirs only between September and mid-March (Hartert & Jourdain 1920) and it is likely that the reservoirs have benefited from an expansion of the breeding range of this species into south-east England (Gibbons et al 1993).

Grey Wagtails have nested in most recent years on the Grand Union Canal and the Wendover Arm. Birds are also present throughout the winter in small numbers and this species can therefore be seen in all months, at any of the four waters. The highest counts are usually during August and September when locally bred juveniles are in evidence.

Pied Wagtail *Motacilla alba*

Breeding resident and passage migrant. Pied Wagtails can commonly be seen at any of the four reservoirs and particularly during the winter and migration periods. Autumn passage is generally heavier than spring with counts of up to 100 birds sometimes recorded. A few pairs breed each year, mostly in local farm buildings and the pale juveniles are a regular autumn feature at the reservoirs.

The White Wagtail *M. a. alba* occurs almost annually, with up to three birds on some days. All but one of the most recent spring records have been during the middle two weeks of April. A male, apparently of this race, held territory in the area of the smallholding adjacent to Wilstone Reservoir during most of April and May 1994, raising the possibility that it was the father of a brood of *alba* wagtails raised there. White Wagtails have been identified in autumn, for example five at the sewage farm on 8 October 1987, although this race probably passes through largely undetected at that season.

*Waxwing *Bombycilla garrulus*

Rare vagrant, four records.

1921:	February
1944:	14, 8 March
1947:	January
1970:	two, flying west, 31 December

These records are predated by an unsubstantiated record in March 1883 (Hartert & Jourdain 1920).

Wren *Troglodytes troglodytes*

Abundant resident. Wrens are present in all suitable habitat along the hedgerows and in the wooded areas and, although not always easy to see, they are the most numerous of the spring songsters. It is likely that this has not always been the case as this species is notoriously badly affected by hard weather such as that in the winter of 1962/3, when the population was judged to have reduced by three-quarters, and 1981/2 when it fell by about one half (Gladwin & Sage 1984). In moderately cold winters, ringing returns suggest that adult survival is much better than that of first-winter birds (Taylor pers comm).

The only recent surveys in the Marsworth/Startops/Tringford Reservoirs and the Wilstone Reservoir areas during May 1994 produced counts of 22 and 39 singing males respectively.

Dunnock *Prunella modularis* (A)

Common resident. Birds of hedgerow and woodland edge, Dunnocks are often inconspicuous around the reservoirs with birds spending most of their time within the ample cover available.

The only recent survey found 12 singing males in the Marsworth/Startops/Tringford Reservoirs area in May 1994 with nine at Wilstone Reservoir. Ringing data show that the true breeding numbers of this species are far greater than these figures might suggest.

Robin *Erithacus rubecula*

Common resident. Robins are generally sedentary in the area and hold territories in all wooded and semi-wooded habitat. Although the national population level is relatively static (Marchant *et al* 1990), ringing data suggest a rise in the population at Tring Reservoirs since 1975.

The four most distant ringing recoveries to date have all been of birds ringed at the reservoirs between June and August, and thus

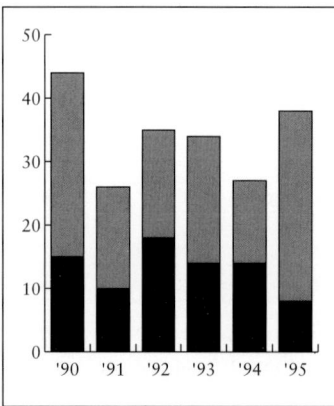

FIGURE 20. DUNNOCK – CES RINGING DATA FOR ADULTS (BLACK BARS) AND JUVENILES (GREY BARS), 1990-95.

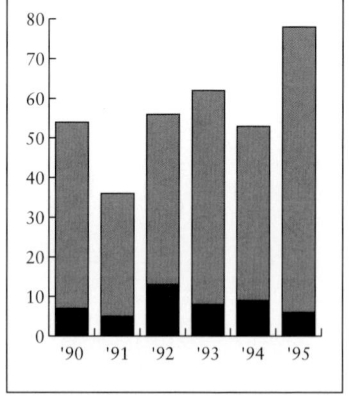

FIGURE 21. ROBIN – CES RINGING DATA FOR ADULTS (BLACK BARS) AND JUVENILES (GREY BARS), 1990-95.

presumably involved locally bred individuals. These birds have dispersed only as far as High Wycombe and Prestwood, Buckinghamshire, Dunstable, Bedfordshire and Bushey, Hertfordshire, with a maximum distance of no more than 28 kilometres.

Nightingale *Luscinia megarhynchos*

Rare passage migrant.

This species nested in the area of the reservoirs until early in the present century, although breeding had apparently ceased by 1920. In recent years it has been a rare and irregular passage migrant with seven spring records since 1974 in the period between late April and early June, none remaining for more than two days. Three autumn individuals have been ringed in the period between 23 and 28 August.

Bluethroat *Luscinia svecica*

Rare vagrant, two records.

1969:	male, 23 September, Marsworth Res
1976:	female, 1 September, Marsworth Res

These occurrences fit well within the typical autumn arrival dates of this largely coastal migrant (Dymond *et al* 1989).

Black Redstart *Phœnicurus ochruros*

Rare passage migrant, six records.

1915:	6 November, Wilstone Res
1942:	28 March, the old refuse dump at Little Tring
1964:	female, 26 April, Wilstone Res
1975:	April
1979:	female, 24 Mar
1984:	18 April, Wilstone Res

Only two of the records have been since 1978 and its relative rarity is particularly surprising given the frequency of occurrence at nearby Blows Down, Bedfordshire, in early spring, occasional breeding at Pitstone cement works, Buckinghamshire, and several sightings in Tring town.

A pair nested successfully at a farm near Tring in the summer of 1979.

Redstart *Phœnicurus phœnicurus*

Scarce passage migrant. A decline in recent years has paralleled the drop in breeding numbers in southern Britain including the local Ashridge Forest population.

There has been a total of 14 records in the period from 1980-94. Records were annual between 1980 and 1982, but there were none in the seven succeeding years. A total of six spring records fall mainly between mid-April and early May with two singles in late June and early July possibly involving wandering failed or non-breeding individuals.

Birds have appeared on autumn passage with extreme dates of 18 July and 23 September, with an exceptionally tardy individual on 26 October 1977. The peak period appears to be in late August and early September, at which time continental migrants may also be involved.

Whinchat *Saxicola rubetra*

Scarce passage migrant. Whinchats are less than annual in occurrence, although a marked decline in the British breeding population has not been reflected in a reduction in the numbers recorded at Tring Reservoirs. Any area of scrub or hedgerow can attract this species, though perhaps the most reliable areas are in the vicinity of the old sewage farm and at Wilstone Reservoir.

Spring passage has produced 30 birds since 1980 with a clear peak between the last week in April and the first week in May accounting for 75 per cent of those recorded. Yearly variation in numbers is shown by the fact that 10 of the birds recorded up to 1994 occurred in a single year, 1989.

The 32 autumn records fall between mid-August and early October with over half in the period between the last week in August and the first week in September. The maximum noted was seven on 3 September 1991.

Stonechat *Saxicola torquata*

Passage migrant and winter visitor. This species is subject to significant periodic population fluctuations related to hard winters but has shown a long term decline for several decades (Marchant *et al* 1990).

In the 1970s wintering numbers within southern Britain were at a relatively high level, but dropped markedly after the winter of 1978/9 and have only recently shown any signs of a recovery. In line with this pattern, it is interesting that there were nine records from 1978-80, followed by only seven more up to 1994.

All recent records have occurred between early October and early April, with the exception of an unseasonal juvenile on 29 July 1978. The only bird to stay more than a couple of days was present from 19 January to 12 February in 1978.

Wheatear *Œnanthe œnanthe*

Scarce passage migrant. Wheatears are significantly more common in spring than in autumn with the numbers recorded in any given year largely dependent on the availability of suitable feeding habitat in the fields adjacent to Wilstone and Marsworth Reservoirs.

In the period between 1980 and 1994 some 62 individuals have been recorded in spring with occurrences in each year, except 1986-88 when only a single bird was noted. Passage covers the period from mid-March to the end of May, but there appear to be two discernible peaks in numbers. The first, presumably involving returning British breeders, is between mid March and mid April and there is a second movement between late April and mid May which probably includes a high proportion of birds destined for Greenland. The peak daily count comprised five males on 11 April 1980.

By contrast the autumn passage is much less noticeable and comprised only 10 individuals in this period. Birds were recorded between mid-August and mid-October but with no clear period of peak passage. An individual was present at Startops End from 11 to 19 November 1976, the latest noted to date.

Ring Ouzel *Turdus torquatus*

Rare spring passage migrant.

Pre-1978:   two, in April, 'in recent years' (Holdsworth *et al* 1978)

1992:   22 April, Wilstone Res, headed off in a north-easterly direction

The species is surprisingly scarce at the reservoirs given the regularity of its appearance at nearby traditional stop-over sites such as Blows Down, Bedfordshire and Steps Hill, Buckinghamshire. Regular coverage of the fields above the dry canal in April might yield more records. Historically the species was recorded in both spring and autumn (Hartert & Jourdain 1920).

Blackbird *Turdus merula* (A)

*Common
resident.*

A ubiquitous bird in suitable habitat around the reservoirs. Ringing returns have indicated a relatively stable population in recent decades with some 659 individuals trapped between 1967 and 1993. There have been few recoveries of note to date.

Peak counts in recent years illustrate an influx of birds during winter with approximately 100 birds on both 3 December 1980 and 6 November 1985.

Fieldfare *Turdus pilaris*

*Common
passage
migrant
and winter
visitor.*

Recent records have been between mid-October and the end of April. In several years there have been no records until November and autumn passage is generally light, although there is a record of 180 on 27 October 1991. The return movements are often more pronounced with a build-up in March.

Winter numbers vary greatly from year to year and are generally highest in cold weather. At least 1000 birds passed over on 3 January 1979 with 400 reported on 16 December 1982.

Song Thrush *Turdus philomelos* (R)

*Resident
and
passage
migrant.*

Often more difficult to see around the reservoirs than Mistle Thrush, the Song Thrush population appears to be following the national trend in declining significantly. Ringing totals have indicated a sharp decline since 1967, with 244 between 1967 and 1975, 189 between 1976 and 1984 and only 109 between 1985 and 1993. Seven singing males were located around the reservoirs during the 1994 survey of singing birds.

Redwing *Turdus iliacus*

*Common
passage
migrant
and winter
visitor.*

The first returning birds generally arrive at the end of September, though a single on 12 September 1992 was particularly early. Passage generally peaks in October and early November, with a high count of 280 on 14 November 1980.

While winter influxes are partly influenced by cold weather, they do not necessarily occur at the same time as Fieldfares, and the highest recent counts have been of 320 on 11 February 1990 and 300+ on 30 December 1993. The last birds linger until early to mid April but most have usually gone a month earlier.

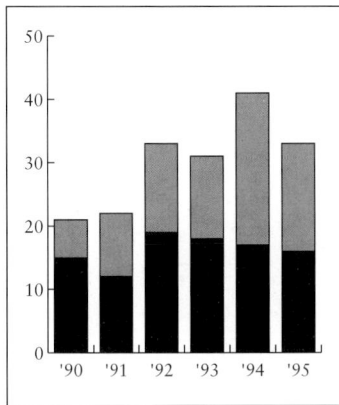

FIGURE 22. BLACKBIRD – CES RINGING DATA
FOR ADULTS (BLACK BARS) AND JUVENILES
(GREY BARS), 1990-95.

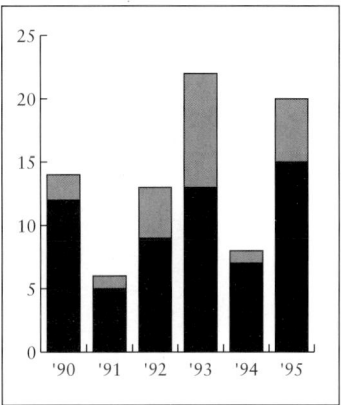

FIGURE 23. SONG THRUSH – CES RINGING
DATA FOR ADULTS (BLACK BARS) AND
JUVENILES (GREY BARS), 1990-95.

Mistle Thrush *Turdus viscivorus*

Common resident, passage migrant and winter visitor.

In the 10 year period from 1978 Mistle Thrushes were rather scarce in the area with peaks of no more than three per day, apart from an influx of up to five birds daily in early 1979 during a spell of hard weather. There has apparently been a marked increase in numbers since then, with up to five breeding pairs in recent years. The species often gathers in flocks in early autumn giving counts of 15 birds in September 1993 and 1994.

*Cetti's Warbler *Cettia cetti* (A)

Rare visitor, seven records.

1977:	male, 22-25 June, Marsworth Res
1977:	female, 1 July, trapped at Wilstone Res, had been previously ringed on Jersey, Channel Islands, on 25 July 1976
1978:	'during October', Wilstone Res
1983:	male, 2 May, Marsworth Res
1985:	1 Jan, sewage farm
1995:	a pair, Wilstone Res, summered and bred
1996:	3 January, Marsworth Res

The early records relate to a period when good numbers were present in south-east England following the initial colonisation of Kent from 1972. This population subsequently crashed in the hard winters in the

1980s but significant numbers remained along the south-west coast of England. It is likely that the expansion of these populations was responsible for the recent records although, as a resident species, it will always be vulnerable to harsh winter weather.

A male was heard singing on 14th April 1995 and was joined by a female less than two weeks later. The male sang until late June and the female remained in the area until at least the autumn. A juvenile at the same site in early August provided the first evidence of breeding at the reservoirs.

Grasshopper Warbler *Locustella nævia*

Rare visitor and migrant. Single Grasshopper Warblers are recorded in most years singing for a few days from rank vegetation around the dry canal or other sites, mainly around Wilstone Reservoir. In the early part of the century this species is known to have nested, with one or two pairs in 1919 at least, but summering has not occurred in recent years.

Birds trapped at Marsworth and Wilstone Reservoirs in May and July 1994 were the first to be caught since 1982.

*Savi's Warbler *Locustella luscinioides*

Rare vagrant, one record. 1989:   14 July, Wilstone Res, trapped

This individual has the distinction of being the first 'retrap' of a British ringed Savi's Warbler in this country. It had previously been ringed at Brandon Marshes, Warwickshire, two months earlier.

*Aquatic Warbler *Acrocephalus paludicola*

Rare vagrant, one record.

1976:   18 September, Wilstone Res

This bird appeared during a particularly marked national influx of this species. Aquatic Warblers are rare inland and more normally recorded from south coast reedbeds.

Sedge Warbler *Acrocephalus schœnobœnus*

Common summer visitor and migrant.

Sedge Warblers breed throughout the area surrounding the reservoirs and along the canals but have been declining in line with national trends (Marchant et al 1990). It is estimated that between 30 to 50 pairs breed, but an intensive survey of a 23 hectare plot around the sewage farm in 1981 produced a count of 39 males. This area has subsequently been 'improved', resulting in a loss of suitable habitat.

Birds generally arrive from the end of March and the last have been reported during early October although, as with all warblers, they become less conspicuous once they stop singing during June.

Large numbers of juveniles pass through the area during August and early September, although the proportion of juveniles to adults has reduced over recent years (see Figure 24).

Several birds ringed at Tring Reservoirs have been reported on migration from the Channel Islands and France although Figure 26 shows that this species may be less reliant than Reed Warbler on the migration route running north-east to south-west along the Chilterns.

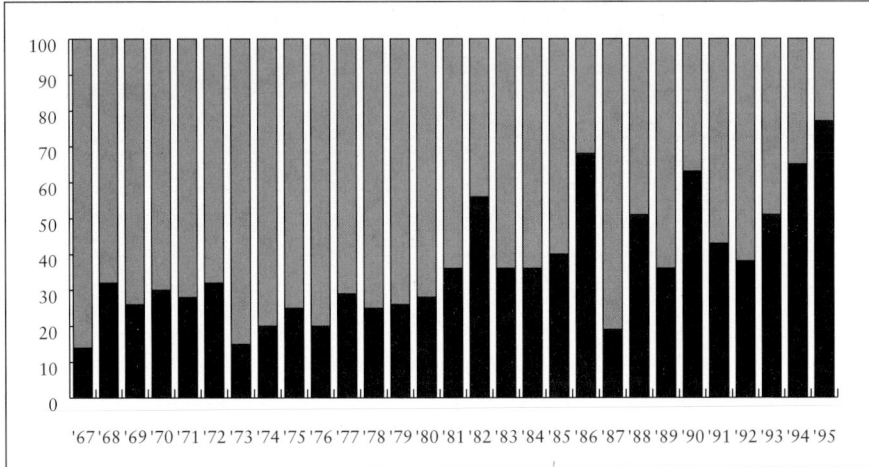

FIGURE 24. THE PERCENTAGE OF ADULT (BLACK BARS) AND JUVENILE (GREY BARS) SEDGE
WARBLERS RINGED IN EACH YEAR, 1967-95.

*Paddyfield Warbler *Acrocephalus agricola*

Rare
vagrant,
one record.

1981:   9 November, photographed near the Drayton Bank hide at
Wilstone Res

This extraordinary occurrence was the first record for Hertfordshire
and the sixth British record of this eastern European and Asian species.
It remains the only inland British record.

THE
PADDYFIELD
WARBLER
PHOTOGRAPHED
NEAR THE HIDE
AT WILSTONE
RESERVOIR ON
9 NOVEMBER
1981.

*Marsh Warbler *Acrocephalus palustris*

Rare vagrant, two records.

1941:   24 June and 4 July, Wilstone Res
1942:   17-18 July

This is a difficult species to identify, even today with advanced knowledge and equipment. The published details, particularly for the second record, are scant. There have only been four recently accepted records in both Hertfordshire and Buckinghamshire.

Reed Warbler *Acrocephalus scirpaceus*

Abundant summer visitor and migrant.

The large reedbeds at Wilstone and Marsworth Reservoirs are an extremely important breeding and feeding areas for this species. Probably over 100 pairs breed with huge numbers using the reeds as a migration staging post in August and September. These have included both adults and juveniles from many sites throughout the country. In the period 1987-95 at the Wilstone Reservoir colony alone, 55 adults and 17 juveniles (first-year birds) have been caught that had previously been ringed elsewhere. The majority of first-year birds have been from the sizeable colony at nearby Weston Turville Reservoir, Buckinghamshire, several of which have returned to breed at Wilstone Reservoir in subsequent years. The timing of recoveries shown in

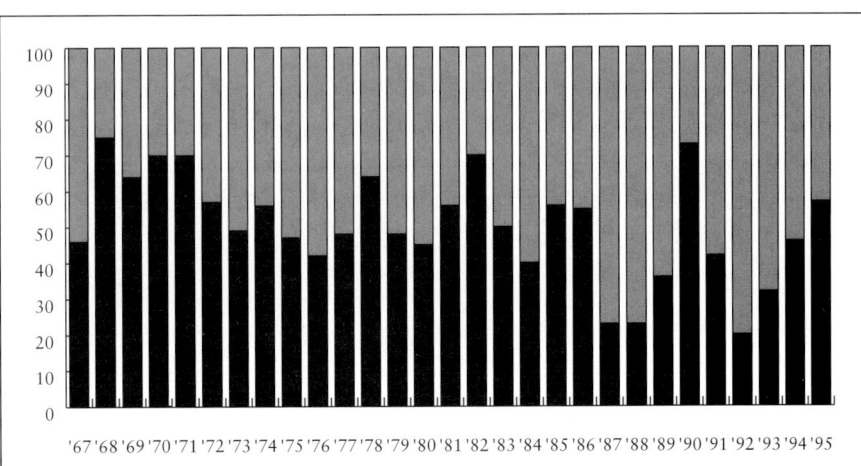

FIGURE 25. THE PERCENTAGE OF ADULT (BLACK BARS) AND JUVENILE (GREY BARS) REED WARBLERS RINGED IN EACH YEAR, 1967-95.

Figure 26 indicates that in spring Reed Warblers pass through coastal marshes at such places as Icklesham and Chichester, East Sussex, moving into the Thames valley and through the Goring Gap. They then follow the Chilterns north before spreading out across the Vale of Aylesbury. In autumn the reverse appears to happen. 1700 birds have been ringed at Tring Reservoirs over the last six years alone.

The Reed Warbler is known to be an important host for Cuckoos.

FIGURE 26. MOVEMENTS WITHIN BRITAIN OF REED AND SEDGE WARBLERS RINGED AND CONTROLLED AT TRING RESERVOIRS, 1967-83 AND 1991-94.

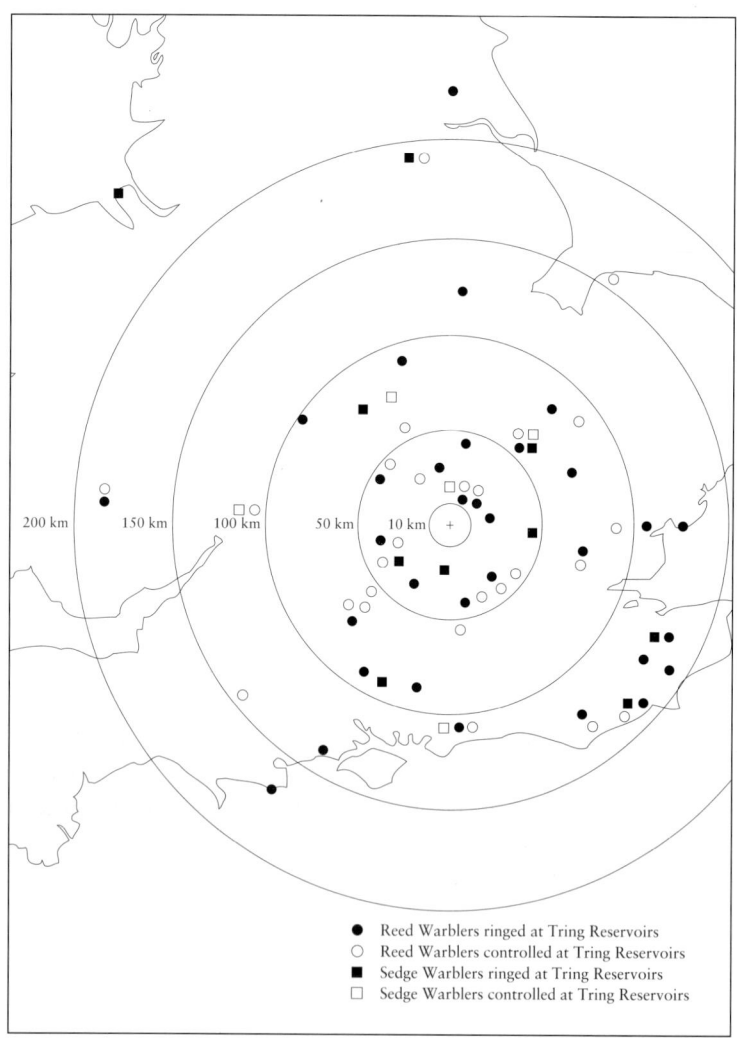

● Reed Warblers ringed at Tring Reservoirs
○ Reed Warblers controlled at Tring Reservoirs
■ Sedge Warblers ringed at Tring Reservoirs
□ Sedge Warblers controlled at Tring Reservoirs

*Great Reed Warbler *Acrocephalus arundinaceus*

Rare vagrant, one record.

1946:   male, 27 April, seen and heard at Marsworth Res

This record constitutes the only record in Hertfordshire and Buckinghamshire, although whether the bird actually frequented the former county is open to some debate.

*Melodious Warbler *Hippolais polyglotta*

Rare vagrant, one record.

1971:   12 September, Marsworth Res

This well watched individual was only the second noted in Hertfordshire. Inland records remain highly unusual.

Lesser Whitethroat *Sylvia curruca*

Common summer visitor.

This species can be found singing from hedgerows around the area from mid-April and occurs until mid-September although there is a record of two birds as late as 31 October.

Birds on return passage move through from late July and this species is often included in the groups of migrating warblers to be found in the bushes during August. A count on 3 August 1989 located 41 birds in the area.

A bird of taller and more dense scrub and hedgerow, the subtle habitat preference of Lesser Whitethroat is shown by the fact that over twice as many have been ringed recently compared to its close relative the Whitethroat which is arguably more numerous in the area.

Whitethroat *Sylvia communis*

Common summer visitor.

The Whitethroat is a very obvious bird when proclaiming territory during the spring. After suffering a major population decline in the late 1960s this species now appears in rather variable numbers with up to 25 singing males around the reservoirs in good years. Birds arrive in mid-April and have been recorded until the last week in September. A count on 3 August 1989 found 32 birds in the Wilstone Reservoir area.

The Whitethroat's preference for open hedgerows has resulted in only 15 having been caught by the ringers in the last six years.

Garden Warbler *Sylvia borin*

Common summer visitor.

With its song easily confused with that of the Blackcap and its subtle plumage, many visitors may overlook this species. Garden Warblers are, however, reasonably common around the reservoirs with pairs breeding in the woodland at Wilstone and Marsworth. Many singing birds choose the thickets adjacent to the dry canal and this is generally the best place to hear them. As with other *Sylvia* warblers, birds pass through in autumn and 17 were counted in the area around Wilstone Reservoir on 3 August 1989.

Due to the more advanced encroachment of scrub into the former reedbed habitat at the ringing site, in recent years there have been 50 per cent more birds ringed at the Marsworth Reservoir CES than that at Wilstone Reservoir.

Blackcap *Sylvia atricapilla*

Common summer visitor and rare winter visitor.

This species is much more numerous than its congener the Garden Warbler. Due to their earlier arrival, Blackcaps often manage to raise a second brood in the scrubland surrounding the reservoirs.

Ringing has shown that there is a significant passage of birds through the site during autumn. In October 1973 a Marsworth bird was reported from Estremadura, central Portugal and in 1982 another was killed at Bignona, Senegal.

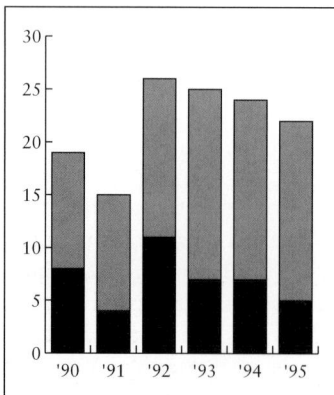

FIGURE 27. GARDEN WARBLER – CES RINGING DATA FOR ADULTS (BLACK BARS) AND JUVENILES (GREY BARS), 1990-95.

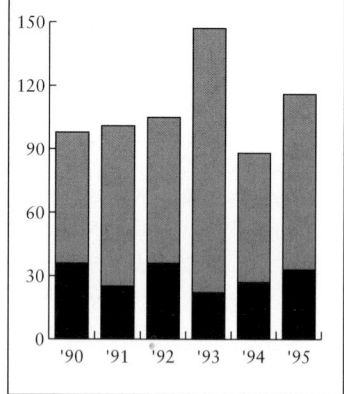

FIGURE 28. BLACKCAP – CES RINGING DATA FOR ADULTS (BLACK BARS) AND JUVENILES (GREY BARS), 1990-95.

A few individuals have been recorded during the winter months in the relatively mild environment around the reservoirs. However, these occurrences have not mirrored the large increase in wintering Chiffchaffs, as they have elsewhere in the region (Dennis 1993; HNHS 1994), and most records of Blackcaps at this time may refer to late or early passage birds en route to wintering areas in the milder climes of south-west England and Ireland.

Wood Warbler *Phylloscopus sibilatrix*

Rare and irregular migrant. Considering the proximity of the former breeding site at Ashridge Forest, Wood Warblers have always been surprisingly rare at the reservoirs. Occurrences have been restricted to passage birds and this is presumably attributable to the lack of suitable habitat. The following have been recorded since 1980:

1983:	7 September
1987:	29 August
1989:	27 April, Marsworth Res
1990:	juvenile, 18 August, Startops End Res
1993:	juvenile, 30 July, Marsworth Res

An exceptional count of three birds was recorded during an unprecedented fall of passerines at the reservoirs on 3 May 1978.

Chiffchaff *Phylloscopus collybita*

Common summer visitor and scarce winter visitor. Chiffchaffs are traditionally the first warbler to appear in spring with migrants generally arriving back at the reservoirs from mid-March. During the summer, Chiffchaffs breed in all the woodland areas around the reservoirs and the majority of birds will have departed by mid-October.

Over the last few years, it has become difficult to distinguish new arrivals from the small but increasing wintering population. This is consistent with other local increases in the south-east of England (Dennis 1993). During the winter Chiffchaffs join the flocks of tits and other small birds and in January 1994 up to five were present around Wilstone Reservoir. Overwintering is not a totally new phenomenon however, Hartert & Jourdain (1920) reported a bird present for a week in mid-February prior to 1920.

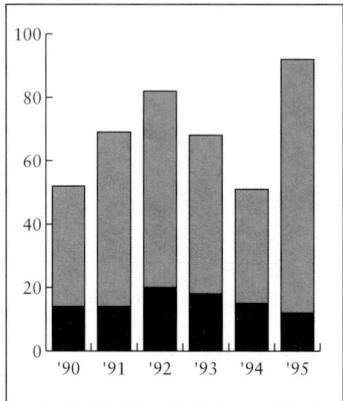

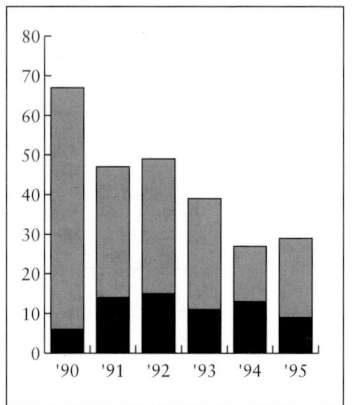

FIGURE 29. CHIFFCHAFF – CES RINGING
DATA FOR ADULTS (BLACK BARS) AND
JUVENILES (GREY BARS), 1990-95.

FIGURE 30. WILLOW WARBLER – CES RINGING
DATA FOR ADULTS (BLACK BARS) AND
JUVENILES (GREY BARS , 1990-95.

Willow Warbler *Phylloscopus trochilus*

Common summer visitor and migrant.

Male Willow Warblers are heard singing from early in April but there is some suggestion that the first birds are migrants passing through and that the local breeding population arrives a little later (Ferguson & Lack 1993). The females appear a couple of weeks after the males. In late summer the adults undergo a complete moult prior to returning to Africa, via the Iberian Peninsula, during late August.

The Willow Warbler was considered by Gibbons *et al* (1993) to be the commonest summer visitor to Britain and birds breed in suitable habitat throughout the Tring Reservoirs area. Numbers have decreased locally in recent years. This is perhaps exacerbated by the maturing of scrubland into less suitable woodland, but there also appears to be a general reduction in both breeding birds and migrants. This species did not suffer the large population reductions noticeable in migrants wintering in the Sahel zone, particularly from the early 1970s. It is still uncertain what factors are responsible for these recent decreases and whether they will prove to be a long-term phenomenon, but ringing data suggest that the overall reduction is the product of a lower number of juveniles (Figure 30).

Goldcrest *Regulus regulus*

Common resident and migrant.

Goldcrests are easiest to find in the autumn and winter when they join roving parties of other small birds. At other times, unless calling, this diminutive species can be hard to locate. Because of their small size, Goldcrests are particularly susceptible to hard weather and, following such extremes in 1986, there were no birds recorded at the reservoirs until mid-January 1988. By 1995 however the population was considered to be at an historic high.

A pair have bred in the woodland at the southern corner of Startops End Reservoir on a number of recent occasions and birds have also nested at Wilstone Reservoir.

During migration periods large numbers pass through the area with groups in excess of 20 not unusual.

Firecrest *Regulus ignicapillus*

Rare migrant.

Firecrest records have been widespread in both migration periods with birds reported between the end of March and the end of April in spring and between the end of September and the end of November in autumn. The two winter records are the only birds to have remained more than one day and include the first occurrence of a pair, on 21 January 1951, one of which remained in the area until 23 March.

Records since 1971 have coincided with the establishment of the breeding colony nearby in Wendover Woods, although breeding birds are usually only present there between the end of April and mid-August. The most recent record was of a single bird during an influx of Goldcrests on 28 November 1992.

Spotted Flycatcher *Muscicapa striata* (R)

Summer visitor and passage migrant.

In line with national and regional trends, this species has become markedly scarcer in recent years although a few pairs still nest annually. Birds have proven to be fairly catholic in their choice of breeding areas, having used the quiet woodland to the south of Tringford Reservoir and the area adjacent to the busy car park at Wilstone Reservoir.

Migrants are more widespread with birds seen at all reservoirs, particularly during the autumn when groups of up to 10 have been found in recent years.

Pied Flycatcher *Ficedula hypoleuca*

Rare migrant. Traditionally thought of as a bird of western Oak woods, the 10 recent records of migrant Pied Flycatchers have occurred in approximately half of the years since 1980. The woodland at the southern end of the Startops End/Marsworth causeway is a particularly favoured site.

Birds occur mainly in autumn between mid-August and the end of September. The rather few spring records fall between the middle of April and early May.

*Bearded Tit *Panurus biarmicus*

Rare autumn and winter visitor. The first record of Bearded Tit at the reservoirs appears to be of a pair shot at Wilstone on 21 December 1848. A number of Dutch birds were released around 1900 but disappeared soon afterwards (Hartert & Jourdain 1920).

Four birds during early 1959 were the start of a remarkable series of records which involved a number of sites in Hertfordshire (Gladwin 1975). Over the next 22 years, annual irruptions resulted in variable numbers of Bearded Tits visiting the reedbeds at Wilstone or Marsworth Reservoirs or the tall herbage at the sewage farm. Birds were recorded mainly during the autumn months, with a maximum count of at least 25 in November 1972. Birds re-trapped in Hertfordshire included some ringed in East Anglia but an individual caught at Marsworth in December 1980, which had previously been ringed at Lodmoor, Dorset, was exceptional. Movements of Tring

A BEARDED TIT IN THE HAND DURING RINGING ACTIVITIES AT WILSTONE RESERVOIR.

Reservoirs birds were recorded to both Essex and Kent.

Since December 1981 the only records involved a series of winter and autumn birds annually between 1985 and 1989, involving up to six individuals, and one on 1 January 1994 at Wilstone Reservoir which was presumably the same bird trapped on 26 March.

A number of these recent records imply that some birds may spend the winter at the site, a proposition not supported by the previous analysis (Gladwin 1975). Occurrences since the mid 1980s echo a decline of this species at local level (Lack & Ferguson 1993; HNHS 1994) and perhaps the recent reduction in the East Anglian population.

Long-tailed Tit *Ægithalos caudatus*

Common breeding resident.

Flocks of Long-tailed Tits can be met throughout the year but are most noticeable during autumn when they combine with roving parties of other small birds.

Although often hidden within dense cover, the intricate nest can sometimes be found during the summer months. No data are available to estimate the breeding population.

Due to their small size, Long-tailed Tits are susceptible to extreme weather and the local population was extinguished by the cold winter of 1917. Interestingly they were not as badly affected as Goldcrest by the bad weather of 1986.

Marsh Tit *Parus palustris* (A)

Scarce breeding resident.

One or two pairs of Marsh Tit breed most years in the private area behind Wilstone Reservoir and breeding probably occurs at Tringford Reservoir.

During the winter they can be easier to locate when they join the wandering parties of commoner tits.

Willow Tit *Parus montanus* (A)

Scarce breeding resident.

Despite its recently reported scarcity in much of the rest of Hertfordshire, this species appears to outnumber Marsh Tits by about five to one in the woods at Wilstone Reservoir (Taylor *in litt*) and are also more numerous than Marsh at Tringford and Marsworth Reservoirs. The reason for this is unclear but it is possible that birds are able to find suitable nesting habitat amongst the derelict woodland containing considerable amounts of rotting timber.

THE WOODLAND BEHIND TRINGFORD RESERVOIR IS A GOOD SITE TO SEE WILLOW TITS.

During winter, this species also joins flocks of tits and other passerines.

Willow Tits tend to be very sedentary as shown by a bird ringed at Tring Reservoirs in May 1966 and recovered at nearby Weston Turville Reservoir in April 1975, a distance of approximately five kilometres. At nearly nine years, this bird was the oldest reported to the national ringing scheme at that time.

Identification from Marsh Tit is always a challenge and distant, non-calling birds are sometimes best left unattributed to either species.

Coal Tit *Parus ater*

Rare visitor. Coal Tits are common in large areas of coniferous habitat within a few kilometres of the reservoirs but recent records average less than one bird per year. Arrivals are concentrated mainly in the autumn months between August and November, with a distinct peak in August. This is in contrast to their visits to other atypical habitats, such as gardens, which are normally visited late in the winter when the birds' natural food becomes scarce. It seems likely that records at the reservoirs relate to some form of post-breeding dispersal. Most occurrences are of single birds and the four birds in October 1991 remained apart at different areas around the reservoirs.

Once present, a Coal Tit can spend some months in the company of a tit flock, as shown by a bird which overwintered during 1991/92.

110

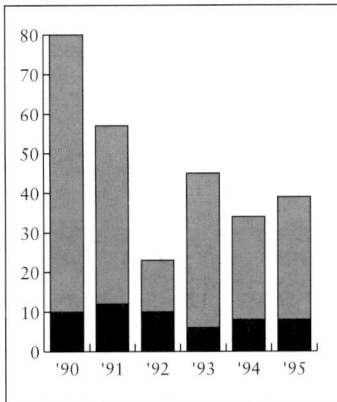

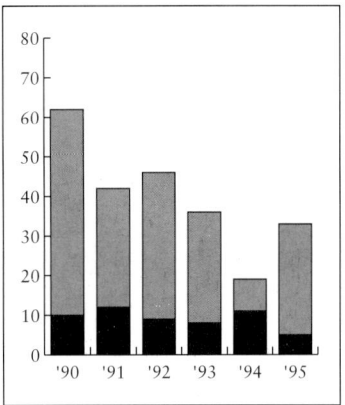

FIGURE 31. BLUE TIT – CES RINGING DATA FOR ADULTS (BLACK BARS) AND JUVENILES (GREY BARS), 1990-95.

FIGURE 32. GREAT TIT – CES RINGING DATA FOR ADULTS (BLACK BARS) AND JUVENILES (GREY BARS), 1990-95.

Blue Tit *Parus cæruleus*

Common breeding resident.

The Blue Tit is ubiquitous around the reservoirs and breeds throughout all the woodland areas. During the winter months, the species makes up a large part of any small passerine flock and many birds feed in the reedbeds.

There has been a reduction in numbers ringed over the past decade but it is not clear if this relates to an actual decrease or to other factors such as habitat changes.

Great Tit *Parus major*

Common breeding resident.

The largest of the titmice, Great Tits are common in all suitable habitats of the area. With its loud and well-known song, the Great Tit is a very obvious species in spring and has made full use of nestboxes provided at certain locations. During the winter, birds often join with other tits and small passerines in feeding flocks.

Nuthatch *Sitta europæa*

Rare visitor, mainly in autumn.

The absence of this species from the areas of mature woodland at the reservoirs is presumably related to the lack of Oak and Elm trees. Autumn records probably refer to young birds wandering from the large populations of Nuthatch that reside in the Chiltern woodlands and in Tring town. Only eight records have been reported since 1980 with two in April and the rest all between the middle of August and the middle of September.

Treecreeper *Certhia familiaris*

Resident, breeding in small numbers.

Although often perceived as requiring many of the same habitat features as the Nuthatch, Treecreepers are able to find suitable conditions at the reservoirs and several pairs breed in the mature and decaying willows and poplars in all the woodlands in the area. During the winter, birds are often found amongst feeding tit flocks.

*Red-backed Shrike *Lanius collurio*

Rare vagrant.

Formerly considered a summer resident, the decline of the Red-backed Shrike was already being documented in 1920 (Hartert & Jourdain 1920). At this time it was commented that only 'about one pair' had nested recently, although further nesting was noted until the 1930s. Occasional migrants were recorded until 1944, but since then there has been only one record, of an immature seen on 23 and 30 August 1977.

*Great Grey Shrike *Lanius excubitor*

Rare winter visitor.

Apart from a period during the 1970s, when the species was scarce but almost regular at the reservoirs with at least four records, the Great Grey Shrike has always been a great local rarity. There were only two records prior to this period and none since the bird which remained at the sewage farm from 13 January 1979 until 3 April, a typically long-staying individual. This appears to reflect a national decline in the number of wintering birds (Fraser & Ryan 1995).

Jay *Garrulus glandarius*

Scarce visitor.

Largely due to the lack of Oak trees in the immediate vicinity, the Jay is the rarest of the crows occurring at the reservoirs. There are sporadic records throughout the year although birds are most regularly recorded between March and May. This species is also scarce in the Vale of Aylesbury, to the west of the reservoirs.

There has been some suggestion that a pair has bred in the woodland at Tringford Reservoir in recent years, although this has not been confirmed. Birds were present all summer at Wilstone Reservoir in 1995 and 1996.

A bird of a continental race was shot in the early 1900s (Hartert & Jourdain 1920).

Magpie *Pica pica*

Common resident.

This controversial species has a healthy, but not large, population at the reservoirs and can usually be found in the farmland around the area. As elsewhere, there is no evidence that this has a significant effect on the populations of other local breeding species.

Winter sees a small influx of Magpies, when the resident population is augmented by birds roosting in the scrub surrounding the reservoirs. There were, for example, 35 at Wilstone Reservoir on 14 December 1980 and 40 birds three months later.

Jackdaw *Corvus monedula*

Resident breeding species.

A few birds remain in the area during the summer months and can be found breeding in small numbers in the woods and farmland around both Tringford and Wilstone Reservoirs.

The large roost near Long Marston begins to attract birds from late August and continues until at least February, with largest numbers occurring during December and January. During this period flocks of mixed crows, largely of this species, can be seen flying north-west over the reservoirs around dusk. Although often difficult to count accurately, studies of this movement have estimated that between 1000 (2 November 1985) and 3000 (1 January 1995) Jackdaws can be involved.

Rook *Corvus frugilegus*

Abundant visitor.

During the autumn and winter months birds join the crow roost nearby and can be seen heading north-west over the reservoirs at dusk. Numbers are considerably less than Jackdaw with no counts exceeding 460 birds (late February 1980). During the summer months, birds visit the farmland in the area from nearby rookeries although there is now no colony in the immediate vicinity.

Over the last 20 years, systematic counts of Rookeries have revealed that the Rook population in Hertfordshire has decreased by 38 per cent, a decline mirrored in the local area covered by the BTO survey square SP91 containing the reservoirs (Dee 1994). A small rookery existed at Tringford Reservoir during the 1970s, with five pairs in

1975 and 11 in 1976, but the colony became extinct by the end of the decade. The formation and subsequent demise of this colony echoes the Hertfordshire trend for larger rookeries to fragment during the 1960s and 70s but to re-consolidate into larger ones over recent years as the population decline continued (Sage 1976; Dee 1994).

Carrion Crow *Corvus corone*

Common breeding resident. Despite being a common and obvious bird, the current numbers of breeding Carrion Crows are unknown, although at least four pairs bred in 1989. As with Jackdaw and Rook, this species is most conspicuous in the autumn and winter when hundreds of birds can be seen flying to the nearby roost with these species. Maximum counts of Carrion Crows in this movement have included 300 on 25 November 1987, and a similar number on 3 October 1992. The former roost at Tringford Reservoir has not been used for many years.

Only six occurrences of the Hooded Crow *C. c. cornix* have been recorded during the past four decades:

1961:	8 March
1966:	6 November, Wilstone Res
1967:	19 March
1975:	22 November, over the sewage farm towards the Tringford roost at dusk
1976:	21-30 March, Tringford roost
1977:	15 December

The lack of recent records parallels the decrease elsewhere in southern and eastern England of this formerly regular winter visitor from Scandinavia. Sage (1959) noted records of singles on about a dozen occasions and commented that it was known locally in the Tring and Berkhamsted areas as the 'Dunstable Crow'.

Starling *Sturnus vulgaris* (A)

Abundant breeding species. This species breeds commonly around the reservoirs but is most conspicuous in autumn and winter when spectacular numbers gather to roost in the reedbed at Wilstone Reservoir. For reasons which are not clear, numbers using the roost fluctuate considerably, but estimates have included 40,000 on 20 December 1992 and 100,000 on 26 December 1993. This roost attracts the attention of local Sparrowhawks on a regular basis.

House Sparrow *Passer domesticus*

Common breeding species.

Breeding numbers of this species are relatively low with the only suitable nesting habitat being the farm buildings adjacent to Startops End and Wilstone Reservoirs and the cottages along the Grand Union Canal. Feeding parties can be seen along the banks of all of the reservoirs but they are most numerous around habitation, with few suitable areas of stubble in recent years.

Tree Sparrow *Passer montanus* (R)

Rare and decreasing winter visitor.

The national population of Tree Sparrow was described by Marchant *et al* (1990) as being strongly in decline since 1976-77 and this situation is echoed at Tring Reservoirs. There are now no breeding pairs in the area, but prior to the mid 1970s breeding was noted at both Tringford and Wilstone Reservoirs and, until the 1980s, there was a colony of over 50 pairs nearby at Miswell Farm (Clark pers comm). Their disappearance from Marsworth Reservoir was noted in 1980 when breeding had also stopped at New Mill, where up to four pairs had nested annually (Spencer *in litt*).

Flocks of up to 200 occurred in winter in the 1970s. Recent records relate to birds around the smallholding at Wilstone Reservoir during November and December 1990 where up to five birds spent several weeks. A bird on the frozen reservoir in February 1991 was presumably one of these birds. Subsequently there have only been four records, from land along the Wendover Arm with between 12 birds (November 1991) and one bird (March 1993).

The last Tree Sparrow ringed, of a total of 116, was in 1979.

Chaffinch *Fringilla cœlebs*

Abundant breeding species and winter visitor.

The most numerous and widespread finch around the reservoirs, Chaffinches can be found in good numbers throughout the year. During spring and summer males proclaim territory from exposed perches while the females in their drabber plumage are less conspicuous but commonly seen. A count in the Marsworth/Startops/Tringford Reservoirs area on 23 May 1993 produced 22 singing males which gives some indication of the breeding population in this part of the area.

Winter sees the presence of finch flocks in the fields and vegetation

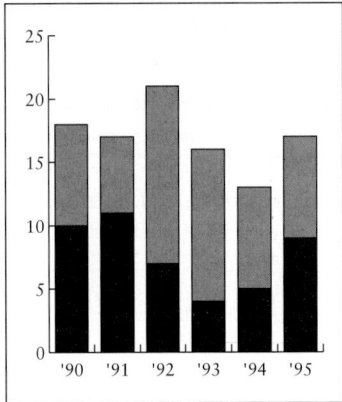

FIGURE 33. CHAFFINCH – CES RINGING DATA
FOR ADULTS (BLACK BARS) AND JUVENILES
(GREY BARS), 1990-95.

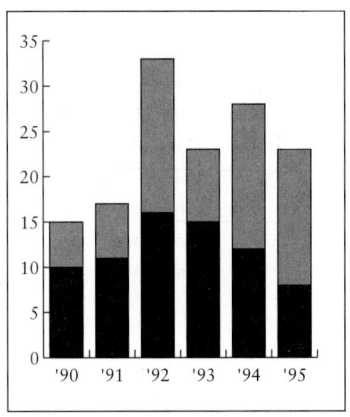

FIGURE 34. BULLFINCH – CES RINGING DATA
FOR ADULTS (BLACK BARS) AND JUVENILES
(GREY BARS , 1990-95.

surrounding the reservoirs, with this species by far the most numerous. Numbers can be quite large with, for example, 250 at Wilstone Reservoir on 25 November 1989.

Brambling *Fringilla montifringilla*

Scarce passage migrant and winter visitor.

Bramblings have been seen at the reservoirs between late October and late April in recent years but birds have mainly occurred during the last two months of this period whilst on spring migration. They can occasionally be found with finch and bunting flocks, but most stay for no more than a day or so. There are only a few records of birds spending extended periods at the site during winter although there is an old record of approximately 300 birds in January and February 1932.

Greenfinch *Carduelis chloris*

Common resident.

Despite the fact that the Greenfinch is a relatively common bird around the reservoirs there are few data to indicate its true breeding status, although it was thought that at least 6 pairs bred in 1989.

Late autumn and winter has produced some sizeable counts, as birds join large finch flocks feeding on open fields. Flocks such as 85 at Wilstone Reservoir on 19 November 1989 have been recorded.

Goldfinch *Carduelis carduelis* (A)

Common resident and winter visitor.

With its bright plumage and light undulating flight the Goldfinch is a popular and familiar sight around the reservoirs. Small numbers breed annually but birds are fairly elusive during this season and it is not until the autumn that flocks start to congregate. Large numbers can then be seen, particularly in late autumn and winter when birds often feed on or close to ground level. Patches of thistles are particularly favoured and can attract flocks such as 85 at Wilstone Reservoir on 19 November 1989 and 100 there on 11 October 1992.

Siskin *Carduelis spinus*

Scarce but regular passage migrant and winter visitor.

Over the past decade Siskins have become noticeably more common and are now reported sporadically during the period between the end of October and the end of March. Birds are most often found feeding in areas of Alders, such as those at Wilstone and particularly Tringford Reservoirs, where numbers of this attractive finch rarely reach double figures, although the largest recent count was of 29 birds in February 1992. The general increase is in line with the national trend as this species has benefited from coniferous afforestation.

Records occurring in September, early October and April, often of birds flying overhead, presumably relate to migrants.

Linnet *Carduelis cannabina* (R)

Common resident and winter visitor.

Small numbers of Linnets breed along the decreasing number of hedgerows which cross the farmland in the area.

During the autumn and winter, birds join together to form flocks which can be found feeding in any areas of weeds and stubble. Numbers in these flocks appear to be reducing, for example a flock of 40-50 birds remained at Wilstone Reservoir throughout much of the autumn and winter of 1989/90 and even larger gatherings were reported in the past. More recently flocks have not exceeded 10 birds and no flock has remained at the reservoirs for an extended period during the last few winters. This decline is thought to be largely a result of changing agricultural practices, with the planting of crops such as winter wheat and the removal of weedy fields and stubble (Gibbons *et al* 1993).

*Twite *Carduelis flavirostris*

*Rare
vagrant,
four
records.*

1953:	two, 15 November, Startops End Res
1961:	29 October, Wilstone Res
1965:	27 December, Wilstone Res
1967:	three or four, 22 January, Wilstone Res

There are only about 20 other records in Hertfordshire and Buckinghamshire.

Redpoll *Carduelis flammea*

*Scarce
migrant
and winter
visitor.*

Redpolls are most often seen feeding in the Alders in the company of Siskins at Tringford Reservoir or in flight over one of the reservoirs. Most records occur between September and April although this species has been recorded in most months. A flock of 25 was seen at Wilstone Reservoir on 14 April 1994, the largest recent count.

While trends are difficult to establish, the species would appear to have declined since the late 1970s when the regional population was at an historic high level (Smith *et al* 1993). The last bird ringed was in 1978, at the height of the population index. Singing birds have been recorded during April and May at both Marsworth and Wilstone Reservoirs, although breeding has never been confirmed.

'Mealy' Redpoll *C. f. flammea* has been recorded at Marsworth Reservoir on 22 January 1922 and at Startops End on 9 December 1923 while, more recently, a flock at Tringford Reservoir in February 1989 included two birds showing the characteristics of this race.

Crossbill *Loxia curvirostra*

*Rare
visitor.*

There are surprisingly few records for the reservoirs despite the proximity of Wendover Woods, where breeding has occurred after the species' periodic eruptions from the continent. Most sightings are of birds overflying the area, although there is one record of three birds alighting briefly.

The following have occurred since 1980:

1988:	two, 28 December, over Wilstone Res
1989:	male and two females, 17 March, landed at Wilstone Res
1989:	nine, 2 October, headed east over Wilstone Res
1991:	two, 7 July, over Wilstone Res

Bullfinch *Pyrrhula pyrrhula* (R)

Resident but decreasing species.

Bullfinches are resident in the area but, due to their liking for dense cover and habit of not forming flocks outside the breeding season, are often one of the hardest regular finches to locate. They frequent areas of scrub such as the dry canal above Wilstone Reservoir and careful investigation can usually locate birds here. As is usual for this species, pairs of birds are often found together.

The breeding status at the reservoirs is unknown and it is difficult to establish if numbers are reducing in line with the national and local decline, which has been particularly marked in farmland areas (Lack & Ferguson 1993, Smith *et al* 1993).

Hawfinch *Coccothraustes coccothraustes*

Rare vagrant.

The British population of Hawfinch appears to be fairly sedentary. Despite the proximity of a well known local Hawfinch site at Ashridge Forest, this species is very rare at the reservoirs, presumably due to the lack of Hornbeam, their favoured foodplant (there is only one in the whole reservoirs area). Records may refer to wandering local individuals during poor food years or possibly immigrants from Europe. There have been four records since 1980:

1983:	16 January, over Tringford Res
1984:	6 April, Wilstone Res
1989:	two, 15 January, Tringford Res
1992:	24 April, over Marsworth Res

Described by Gladwin & Sage (1986) as being a local, fairly common resident, the breeding population in west Hertfordshire appears to be in decline (Smith *et al* 1993) and occurrences of this species may become even less frequent in future.

*Snow Bunting *Plectrophenax nivalis*

Rare vagrant, nine records.

There have been nine records, the first in 1895 when there was a 'large flock' noted near the reservoirs. Other records all occurred in the period between early October and early November with the exception of the only record since 1980:

1981:	8 January, Wilstone Res

Snow Buntings are very scarce inland in south-east England with a
total of approximately 40 records in Buckinghamshire and
Hertfordshire.

Yellowhammer *Emberiza citrinella*

*Common
resident
and winter
visitor.*
Yellowhammers seem to be more successful than many other farmland
species in surviving alongside modern agricultural practices, possibly
because of their ability to capitalise on increased grain availability as
cereal acreage increases. The smaller finches are more reliant on weed
seed which has been in decline as the intensity of herbicide usage
increases (Marchant *et al* 1990).

The current breeding population appears to be fairly stable, with up
to five singing males present particularly along the dry canal and
around Wilstone Reservoir.

Winter flock numbers can be large if stubble fields are available to
them for feeding, with flocks of over 100 recorded at Wilstone
Reservoir on three recent occasions: 22 February 1987, 7 November
1989, and 1-2 January 1993.

*Cirl Bunting *Emberiza cirlus*

*Rare
vagrant.*
This species has always been one of Hertfordshire and
Buckinghamshire's rarest breeding species and there have been few
sightings at the reservoirs since 1945, reflecting its national decline
(Gibbons *et al* 1993). Most birds occurred during winter and included
some relatively large flocks such as 10 at Wilstone Reservoir on
4 February 1941. There was at least one nest found locally, at Drayton
Beauchamp, Buckinghamshire, during 1864.

There has been a long-term decline in this species as noted 'by their
disappearance from Tring since 1917' (Hartert & Jourdain 1920). It
seems very unlikely that the local status of the Cirl Bunting will change
in the near future.

Reed Bunting *Emberiza schœniclus* (R)

*Common
resident.*
The reedbeds at Marsworth and Wilstone Reservoirs harbour a healthy
breeding population of Reed Buntings and several pairs also nest in
areas of rank vegetation and hedges surrounding the reservoirs area.
During 1981, 30 pairs were found on a 23 hectare study site at the
sewage farm, although this had reduced to five pairs by 1984 after the

modernisation of the plant. CES data suggest that the population, and breeding success, can fluctuate considerably between years (Figure 35).

The reedbeds also support a winter roost of this species, and have included some relatively large flocks; for example approximately 200 at Wilstone Reservoir on 13 February 1993.

Corn Bunting *Miliaria calandra* (R)

Scarce resident and local winter visitor. Despite a recent national and local decline in breeding numbers (Dee & Smith 1994), Corn Buntings currently manage to maintain a presence in the reservoirs area during the summer. One or two males have been found singing from prominent hedgerow or overhead wires in the vicinity of James Farm in recent years but it is not clear how long the local population can survive given the species' rapidly declining fortunes.

As a winter visitor Corn Buntings are much easier to find, since a large roost gathers in the Marsworth Reservoir reedbed. Numbers have been as high as 450, usually between late October and March, with the birds congregating in the adjacent trees before dropping into the reeds. This roost held a recent high count of over 200 birds in February 1996 and has reduced noticeably in recent years.

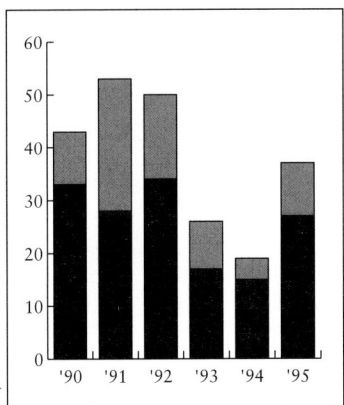

FIGURE 35. REED BUNTING – CES RINGING DATA FOR ADULTS (BLACK BARS) AND JUVENILES (GREY BARS), 1990-95.

Appendix 1 –
Birds believed to originate from captivity

While there is often no direct evidence of captive origin with some of the following records, sufficient doubt exists, however, for them not to be generally considered to relate to wild birds.

Sacred Ibis *Threskiornis æthiopicus*
One on 28 September 1989 until March 1990.

African Spoonbill *Platalea alba*
One on 28-29 September 1991.

White-faced Whistling Duck *Dendrocygna viduata*
One record in May 1988.

Black Swan *Cygnus atratus*
Several records since 1937, with individuals often remaining for long periods.

Coscoroba Swan *Coscoroba coscoroba*
One record, 17 May-mid-November 1935.

Bean Goose *Anser fabalis*
One record early 1985.

Bar-headed Goose *Anser indicus*
Several records.

Canada Goose *Branta canadensis*
Birds of one of the small races have been seen on occasions, for example at Wilstone Reservoir on 1 May 1985, 25 February and 28 March 1987.

Snow Goose *Anser cærulescens*
Several records. A blue phase bird was present on 5 April 1978.

Barnacle Goose *Branta leucopsis*
Rare visitor of unknown origin, probably escapes.
　　Whilst most, if not all, records of this species refer to feral birds it is possible that wild birds could arrive at the reservoirs on occasions. All but two records fall in the periods March-May or September-October.

Cape Shelduck *Tadorna cana*
Recorded regularly. Beware of this species when identifying Ruddy Shelduck.

Ruddy Shelduck *Tadorna ferruginea*
Probably the most common wildfowl not thought to be of wild stock, although at least one record occurred during an influx into Britain.

Chiloe Wigeon *Anas sibilatrix*
Relatively numerous in recent years, due to its popularity in collections, with up to three birds occurring together; beware of this species, or hybrids with Wigeon, when American Wigeon is suspected.

American Wigeon *Anas americana*
There was a male, and possibly a female, seen at the reservoirs on 5 April 1925 which were considered to be escapes.

Cape Teal *Anas capensis*
One at Marsworth Reservoir on 18 April 1995.

Chestnut Teal *Anas castanea*
One record in June 1989.

Bahama Pintail *Anas bahamensis*
An occasional visitor with a number of records.

Silver Teal *Anas versicolor*
Two first-winter birds remained at Wilstone Reservoir throughout their stay from 31 August -21 September 1991.

Cinnamon Teal *Anas cyanoptera*
A single record of a drake at the sewage farm on 25 April 1984.

Rosybill *Netta peposaca*
A male was resident at the reservoirs from November 1992 until July 1993.

Baer's Pochard *Aythya baeri*
A record of a bird shot on 5 November 1901 was the sole accepted British record until reviewed and considered to be an escape from captivity.

Golden Pheasant *Chrysolophus pictus*
One recorded in September 1965 (Holdsworth *et al* 1978).

Sarus Crane *Grus antigone*
One on 4 April 1937.

Grey-headed Gull *Larus cirrocephalus*
A single record of an adult bird on 10 February 1991 at Wilstone Reservoir.

   All birds in captivity in Britain were accounted for at the time and contrary to Clement & Gantlett (1993) no birds were kept at Whipsnade Zoo. The hard weather during February had resulted in significant movements of gulls and the bird was relocated at the Brogborough gull roost, in Bedfordshire, a week later.

Chestnut Munia *Lonchura malacca*
In 1898 a 'small flock' of this species was seen in the reedbeds at Tring Reservoirs, one of which was shot (Sage 1959).

Bronze Manikin *Lonchura cucullata*
A single bird at Wilstone Reservoir on 20 June 1996.

The wildfowl collection recently established at the smallholding adjacent to Wilstone Reservoir includes a number of pinioned species that may be seen as escapes at the reservoirs in the future, for example Barnacle, Bar-headed and Red-breasted Geese and Wood Duck.

Appendix 2 –
First and last dates of regular migrants

Passage migrants	First spring	Last spring	First autumn	Last autumn
Marsh Harrier	20/04	13/05	01/08	27/10
Osprey	06/04	30/05	10/08	30/09
Black-tailed Godwit	30/03	23/04	05/07	28/10
Bar-tailed Godwit	01/05	06/05	23/07	10/10
Whimbrel	16/04	22/05	06/07	09/09
Greenshank	04/03	29/05	26/07	07/11
Little Gull	23/03	01/06	28/08	01/11
Sandwich Tern	26/03	05/05	22/08	25/09
Arctic Tern	08/04	09/06	03/08	20/10
Little Tern	12/04	08/06	05/08	21/09
Redstart	17/04	23/05	18/07	26/10
Whinchat	08/04	22/05	16/08	03/10
Wheatear	16/03	31/05	15/08	16/10
Pied Flycatcher	17/04	02/05	16/08	26/09

Summer visitors	First arrival	Last departure
Hobby	15/04	14/10
Little Ringed Plover	12/03	10/09
Common Tern	04/04	01/10
Cuckoo	08/04	06/09
Swift	21/04	01/10
Sand Martin	07/03	06/11
Swallow	25/03	14/11
House Martin	23/03	27/10
Yellow Wagtail	23/03	18/10
Sedge Warbler	16/03	02/10
Reed Warbler	13/04	27/10
Lesser Whitethroat	20/04	18/09
Whitethroat	13/04	24/09
Garden Warbler	17/04	01/10
Blackcap	24/03	24/10
Willow Warbler	20/03	02/10
Spotted Flycatcher	01/05	25/09

Winter visitors	First arrival	Last departure
Red-necked Grebe	11/09	26/04
Slavonian Grebe	20/10	28/04
Bewick's Swan	11/10	19/05
Smew	08/12	08/03
Golden Plover	23/08	30/05
Jack Snipe	23/09	17/04
Stonechat	09/10	05/04
Fieldfare	13/10	30/04
Redwing	12/09	19/04
Brambling	20/10	27/04
Siskin	23/09	19/04

GOLDENEYES AND A DRAKE GARGANEY.

JPPW

Appendix 3 –
Scientific names of species mentioned in the text

Plants:
Celery-leaved Buttercup *Ranunculus sceleratus*
Rigid Hornwort *Ceratophyllum demersum*
Common Lime *Tilia vulgaris*
Sycamore *Acer pseudoplatanus*
Meadowsweet *Filipendula ulmaria*
Blackthorn *Prunus spinosa*
Hawthorn *Cratægus monogyna*
Spiked Water-milfoil *Myriophyllum spicatum*
Elm *Ulmus procera*
Alder *Alnus glutinosa*
Hornbeam *Carpinus betulus*
Oak *Quercus robur*
Horse Chestnut *Æsculus hippocastanum*
Poplar *Populus sp*
Osier *Salix viminalis*
Bog Pimpernel *Anagallis tenella*
Ash *Fraxinus excelsior*
Mudwort *Limosella aquatica*
Fen Bedstraw *Galium uliginosum*
Elder *Sambucus nigra*
Common Fleabane *Pulicaria dysenterica*
Canadian Waterweed *Elodea canadensis*
Nuttall's Pondweed *Elodea nuttallii*
Hard Rush *Juncus inflexus*
Round-fruited Rush *Juncus compressus*
Blunt-flowered Rush *Juncus subnodulosus*
Green-flowered Helleborine *Epipactis phyllanthes*
Southern Marsh Orchid *Dactylorhiza prætermissa*
Early Marsh Orchid *Dactylorhiza incarnata*
Bulrush *Typha latifolia*
Lesser Bulrush *Typha angustifolia*
Distant Sedge *Carex distans*
Common Reed *Phragmites australis*
Reed Sweet-grass *Glyceria maxima*
Orange Foxtail *Alopecurus æqualis*

Insects:
Emerald Damselfly *Lestes sponsa*
Emperor Dragonfly *Anax imperator*
Brown Hawker *Æshna grandis*
Southern Hawker *Æshna cyanea*
Migrant Hawker *Æshna mixta*
Black-tailed Skimmer *Orthetrum cancellatum*
Four-spotted Chaser *Libellula quadrimaculata*
Broad-bodied Chaser *Libellula depressa*
Common Darter *Sympetrum striolatum*
Yellow-winged Darter *Sympetrum flaveolum*
Clouded Yellow *Colias croceus*
Comma *Polygonia c-album*
Red Admiral *Vanessa atalanta*
Marbled White *Melanargia galathea*
Ringlet *Aphantopus hyperantus*
Speckled Wood *Pararge ægeria*
Holly Blue *Celastrina argiolus*
Essex Skipper *Thymelicus lineola*
Monarch *Danaus plexippus*

Fish:
Pike *Esox lucius*

Birds:
Red-breasted Goose *Branta ruficollis*
Wood Duck *Aix sponsa*
Rough-legged Buzzard *Buteo lagopus*

Mammals:
Noctule *Nyctalus noctula*
Pipistrelle *Pipistrellus pipistrellus*
Daubenton's Bat *Myotis daubentonii*
Brandt's Bat *Myotis brandtii*
Muntjac *Muntiacus reevesi*
Mink *Mustela vison*
Fox *Vulpes vulpes*
Rabbit *Oryctolagus cuniculus*

Appendix 4 –
Contact addresses

Friends of Tring Reservoirs and Herts Bird Club Recorder:
Rob Young, 'Falcon House', 28 Tring Road, Long Marston, Tring,
Hertfordshire, HP23 4QL.

Buckinghamshire Bird Club Recorder:
Andy Harding, 15 Jubilee Terrace, Stoney Stratford, Milton Keynes,
Buckinghamshire, MK11 1DU.

Herts Bird Club:
Ted Fletcher, Secretary, Beech House, Aspenden, Hertfordshire,
SG9 9PG.

Hertfordshire Natural History Society:
John Scivyer, Secretary, 9 Hill Rise, Potters Bar, Hertfordshire,
EN6 2RX.

Hertfordshire & Middlesex Wildlife Trust:
Grebe House, St Michael's Street, St Albans, Hertfordshire, AL3 4SN.

Berkshire, Buckinghamshire and Oxfordshire Naturalists' Trust
(BBONT):
3 Church Cowley Road, Rose Hill, Oxford, OX4 3JR.

British Waterways:
Marsworth Junction, Watery Lane, Marsworth, Tring, Hertfordshire,
HP23 4LZ.

British Trust for Ornithology:
The National Centre for Ornithology, The Nunnery, Thetford,
Norfolk, IP24 2PU.

References

BIND R A. 1995. The Bedfordshire Bird Report for 1994. *The Bedfordshire Naturalist* 49 (Part 2). Bedfordshire NHS.

CRAMP S & SIMMONS K E L (eds). 1977-82. *The Birds of the Western Palearctic.* Volumes I-III. Oxford University Press, Oxford.

— (ed). 1985-92. *The Birds of the Western Palearctic.* Volumes IV-VI. Oxford University Press, Oxford.

— & PERRINS CM (eds). 1993. *The Birds of the Western Palearctic.* Volumes VII. Oxford University Press, Oxford.

DAZLEY R A & TRODD P. 1994. *The breeding birds of Bedfordshire.* Bedfordshire Natural History Society.

DEE C W. 1994. The 1993 Hertfordshire Rookery Census. *Trans Herts NHS* 32:173.

— & SMITH K W. 1994. Bunting Surveys in Hertfordshire in 1992 and 1993. *Trans Herts NHS* 32:186.

DENNIS M K. 1993. Wintering Blackcaps and Chiffchaffs in the London Area. *London Bird Report* 57:145.

DEPARTMENT OF THE ENVIRONMENT. 1995. *Biological Recording in the United Kingdom – Present Practice and Future Development.* London.

DEVLIN T R E, JENKINS A R & LLOYD-EVANS L. 1962. Sociable Plover in Hertfordshire. *British Birds* 55:236.

DYMOND J N, FRASER P A & GANTLETT S J M. 1989. *Rare Birds in Britain and Ireland.* T & A D Poyser, Calton.

EASY G. 1994. An update on Skua movement inland from the Wash. *Cambridge Bird Report* 1993. Cambridge Bird Club.

ELTON G S. 1992-96. *Hilfield Park Reservoir Bird Reports,* 1991-1995.

FITTER R, FITTER A & BLAMEY M. 1974. *The Wildflowers of Britain and Europe.* Collins, London.

— FITTER A & ARLOTT N. 1981. *The Complete Guide to British Wildlife.* Collins, London.

FRASER P & RYAN J. 1995. Status of Great Grey Shrike in Britain and Ireland. *British Birds* 88:478.

CLEMENT P & GANTLETT S. 1993. The origin of species. *Birding World* 6:206.

GIBBONS D W, REID J B & CHAPMAN R A. 1993. *The New Atlas of Breeding Birds in Britain and Ireland: 1988-1991.* T & A D Poyser, London.

GLADWIN T W. 1975. Bearded Tits in Hertfordshire since 1959. *Trans Herts NHS* 27:355.

— & SAGE B L. 1986. *The Birds of Hertfordshire.* Castlemead, Ware.

HARTERT E & JOURDAIN F C R. 1920. The Birds of Buckinghamshire and the Tring Reservoirs. *Novitates Zoologicæ* 27:171.

HAYWARD H H S. 1947. The birds of Tring Reservoirs, in Fletcher M, Clive Rouse E and Viney E (Eds), *Records of Buckinghamshire* Vol XV. The Architectural and Archeological Society of the County of Buckingham.

— 1968. The Coot (*Fulica atra* L.) at Tring Reservoirs Hertfordshire. *Trans Herts NHS* 26:256.

HERTFORDSHIRE NATURAL HISTORY SOCIETY. Hertfordshire Bird Reports 1984-94. *Trans Herts NHS.*

HOLDSWORTH M, HUDSON R, MAGEE J D & PRATER T. 1978. *The Birds of Tring Reservoirs.* Herts NHS, Berkhamsted.

KENNEDY A M W CLARK. 1868. *The birds of Berks & Bucks, a contribution to the natural history of the two counties, Eton & London.*

KRAMER D. 1995. Inland spring passage of Arctic Terns in southern Britain. *British Birds* 88:211.

LACK P. 1986. *The Atlas of Wintering Birds in Britain and Ireland.* T & A D Poyser, Calton.

— & FERGUSON D. 1993. *The Birds of Buckinghamshire.* Buckinghamshire Bird Club.

MARCHANT J H, HUDSON R, CARTER S P AND WHITTINGTON P A. 1990. *Population trends in British breeding birds.* British Trust for Ornithology, Tring.

MEAD C & SMITH K W. 1982. *The Hertfordshire Breeding Bird Atlas.* HBBA, Tring.

O'SULLIVAN J. 1977. Report on rare birds in Great Britain in 1976. *British Birds* 70:405.

ROGERS M J. 1978. Report on rare birds in Great Britain in 1977. *British Birds* 71:481.

— 1980. Report on rare birds in Great Britain in 1979. *British Birds* 73:491.

— 1985. Report on rare birds in Great Britain in 1984. *British Birds* 78:529.

— 1987. Report on rare birds in Great Britain in 1986. *British Birds* 80:503.

— 1995. Report on rare birds in Great Britain in 1994. *British Birds* 88:493.

RSPB. 1996. *Birds of conservation concern in the United Kingdom, Channel Islands and Isle of Man.*

SAGE B L. 1959. *A History of the Birds of Hertfordshire.* Barrie & Rockliff, London.

— 1976. The national survey of rookeries, 1975: Hertfordshire Rookeries. *Trans Herts NHS* 27:361.

SAWFORD B. 1987. *Butterflies of Hertfordshire.* Castlemead, Ware.

SMITH F R. 1968. Report on rare birds in Great Britain in 1967. *British Birds* 61:329.

— 1972. Report on rare birds in Great Britain 1971. *British Birds* 65:322.

— 1973. Report on rare birds in Great Britain 1972. *British Birds* 66:331.

SMITH K W, DEE C W, FEARNSIDE J D, FLETCHER E W & SMITH R N. 1993. *The Breeding Birds of Hertfordshire.* Hertfordshire Natural History Society, Potters Bar.

SWAINE C M. 1962. Report on rare birds in Great Britain in 1961. *British Birds* 55:562.

WOLLEY J. 1853. *Ootheca Wolleyana.*

WITHERBY H F. 1934. Arctic Ringed Plover in Hertfordshire. *British Birds* 28:175.

YOUNG R A. 1991. The rise of the Mediterranean Gull in Hertfordshire. *Trans Herts NHS* 31:298.

Friends of Tring Reservoirs

Friends of Tring Reservoirs (FoTR) is a local conservation group formed during the autumn of 1993.

With a membership in excess of 300, FoTR is working in many areas with a number of other organisations with local interests, such as British Waterways, Thames Water, the Hertfordshire & Middlesex Wildlife Trust and the Rothschild Estate, especially through the Water Bailiff.

The aims of FoTR are to conserve and improve nature conservation at the reservoirs and this is addressed by the following objectives:

• Monitoring and recording the environment –
FoTR maintains the bird log initiated by the BTO and ensures that all records are passed to the relevant county recording organisations. A number of initiatives with local and national organisations are undertaken.

• Conserving and/or improving the environment as appropriate –
The organisation played a major part in the recent implementation of improvements to the pathways around the reservoirs including the fencing of sensitive areas and construction of a new bird hide.

• Encouraging wider public awareness of conservation issues –
Members of the public are encouraged to participate in regular guided walks and a programme of school visits is aimed at younger visitors. FoTR publishes a regular newsletter, *Grebe*, highlighting progress and items of interest.

• Promoting a programme of conservation work –
A programme of work days ensures the maintenance and enhancement of facilities

• Encouraging the development of an environmentally sound Management Plan –
FoTR is working with all interested parties to encourage a comprehensive management plan to be developed for the site.

The Hertfordshire Natural History Society

The Hertfordshire Natural History Society was formed in 1875 and since then has collected and published data on the many forms of natural history in the county. The data collected is used in publications such as this book, as part of national recording schemes and for conservation purposes.

The society holds a programme of indoor and field meetings throughout the year, and an annual long field trip to a site of interest elsewhere in the UK.

The Herts Bird Club is a specialist section of the society and collects and uses the wealth of data about our most obvious form of wildlife. Formed in 1971, as the Society's Ornithological Section, the name was changed in 1989. The club holds an annual conference for those interested in the birds of the county, organises surveys of individual species and groups and assists with national survey work for the British Trust for Ornithology. Most importantly, it collects records of both common and less frequent species for publication in the annual *Hertfordshire Bird Report* which also contains reports on ringing in the county, Common Bird Census, Waterways Bird Survey and Breeding Bird Survey results and other items of interest.

Membership of the Society brings with it access to all meetings, the annual Transactions (full members only) and Hertfordshire Bird Report, and the chance to meet others interested in natural history. Additionally, membership gives you the chance to put your observations to good use in helping to conserve our flora and fauna.

Index